Encyclopaedia of
Nuclear Physics

Encyclopaedia of Nuclear Physics

Volume 3

Quantum Physics

Shahzad Ahmad

ANMOL PUBLICATIONS PVT. LTD.
NEW DELHI - 110 002 (INDIA)

ANMOL PUBLICATIONS PVT. LTD.

H.O.: 4374/4B, Ansari Road, Daryaganj
New Delhi - 110 002
Ph.: 23261597, 23278000

B.O.: No. 1015, Ist Main Road, BSK III Stage
III Phase, III Block
Bangalore - 560 085 (India)
Visit us at: www.anmolpublications.com

Encyclopaedia of Nuclear Physics

First Published, 2006

ISBN 81-261-2618-3 (Set)

PRINTED IN INDIA

Published by J.L. Kumar for Anmol Publications Pvt. Ltd., New Delhi - 110 002 and Printed at Mehra Offset Press, Delhi.

Contents

Preface

Science is the sermon of today. In every walk of life, since childhood till old age, science plays an important role in the life of man. Uptill some decades back, studied merely as a subject in schools and colleges, Science has assumed the status of an essential part of life today. With the expansion of literacy among the masses and growing awareness, they have woken up from the slumber of ignorance and have started taking up things more sincerely and thoughtfully. It is one of the reasons that our nation is heading towards new goals day after day.

It is a vast subject to study and talk about. For the purpose of ease, Science has been divided into more than two hundred and fifty branches. Specialisation in any branch can be obtained at different institutions and universities, scattered all over the country and abroad.

Apart from other super-specific branches, Physics, Chemistry, Zoology, Botany, Mathematics and Computronics, etc. are more general and closer to daily human life. From the point of view of the cut-throat competition in educational world, a thorough study of the above-mentioned branches, proves very handy, for the new aspirants, especially and for teachers and tutors, generally.

An urgent need was being felt for such a series of books, encompassing the commonly-studied and discussed branches of Science of equal significance and presented in an easy language, understood by all. With God's grace, our dream came true and we are happy, while introducing the dream-series for the lovers of knowledge.

The series consists of many a title, covering various branches of Science, such as Physics, Chemistry, Zoology, Botany, Geology, Agriculture Science, Computer Science, Mathematics and so on. The series is important and a must not only for students but also for teachers. Explained in simple language and aided with graphics and diagrams, everything, between the two ends—title and back cover—of the series, has been presented in a perfect way.

This set of books is devoted to **Nuclear Physics**, which comprises three volumes: (1) Nuclear Physics, (2) Atomic Physics and (3) Quantum Physics.

It is hoped that the series would serve both, the students community and the academics, equally and will prove to be a boon for those, aspiring for newer heights. Nevertheless, any suggestions for further betterment of the endeavour will be highly appreciated and solicited.

Editor

1

Quantum Physics

Duality of Wave-Particle

(de-Broglie Wavelength) : Each material particle in motion behaves as waves and the wavelength λ associated with any moving particle of momentum p is given by

$$\lambda = \frac{h}{p} = \frac{h}{mv}$$

- de-Broglie wavelength of particle of K.E. = E_k is given by

$$\lambda = \frac{h}{\sqrt{2mE_k}}$$

- de-Broglie wavelength for charged particle of charge q and accelerated through a potential difference of V volts is

$$\lambda = \frac{h}{\sqrt{2mqV}}$$

- de-Broglie wavelength of a material particle at temperature T is

$$\lambda = \frac{h}{\sqrt{2mkT}} \qquad k \rightarrow \text{Boltzmann constant}$$

- de-Broglie wavelength of helium atom

$$\lambda_{He} = \frac{h}{\sqrt{3mkT}}$$

- The relativistic expression for the de-Broglie wavelength of an electron accelerated through a high p.d. of V volts

$$\lambda = \frac{h}{\sqrt{2m_0 eV}}\left(1 + \frac{eV}{2m_0 c^2}\right)$$

Heisenberg's Uncertainty Principle

- The product of uncertainties in determining the position and momentum of a particle at the same instant is

$$\Delta x \, \Delta p \geq \hbar$$

- If ΔE is the uncertainty in determining energy of the system and $\Delta t \rightarrow$ uncertainty in time

$$\Delta E \, \Delta t \geq \hbar$$

- Also $\Delta J \, \Delta\theta \geq \hbar$

$\Delta J \rightarrow$ Uncertainty in angular momentum

$\Delta\theta \rightarrow$ Uncertainty in angle

Exact Statement : The product of the uncertainties in determining the position and momentum of the particle can never be smaller than the number of the order $\frac{\hbar}{2}$

$$\therefore \quad \Delta x \, \Delta p \geq \frac{\hbar}{2}$$

Similarly, $$\Delta E \, \Delta t \geq \frac{\hbar}{2}$$

$$\Delta J \, \Delta\theta \geq \frac{\hbar}{2}$$

Applications:

(i) Non-existence of the electrons in the nucleus.

(ii) Radius of Bohr's first orbit.

(iii) Minimum energy of a harmonic oscillator.

(iv) Energy of a particle in one dimensional box.

(v) Finite width of spectral lines.

Schrodinger Wave Equation

(i) Time independent Schrodinger equation

$$\nabla^2\psi + \frac{2m}{\hbar^2}(E - V)\psi = 0$$

(ii) Time dependent Schrodinger equation

$$\left(-\frac{\hbar^2}{2m}\nabla^2 + V\right)\psi = i\hbar\frac{\partial\psi}{\partial t}$$

The operator $\left(-\frac{\hbar^2}{2m}\nabla^2 + \dot{V}\right)$ is called Hamiltonian represented by H $\Rightarrow$ while operator $i\hbar\frac{\partial}{\partial t}$, operated on ψ, gives E

$\therefore$ TDSE $\Rightarrow$ $H\psi = E\psi$

The Commutator : The commutation relation can be written as

$$[xp_x - p_x x] = [x, p_x] = i\hbar$$

The quantity

$$[\alpha, \beta] = \alpha\beta - \beta\alpha = -[\beta, \alpha]$$

is said to be the commutator of the two operators α and β.

Probability Current Density

$$\bar{J} = \frac{i\hbar^2}{2m}[\psi\nabla\psi^* - \psi^*\nabla\psi]$$

and $\psi^* \psi \rightarrow$ Position probability density.

- *Expectation values*

$$< x > = \frac{\iiint \psi^* x \, \psi d\tau}{\iiint \psi^* \psi d\tau}$$

Since $\iiint \psi^* \psi d\tau = 1$

$$\therefore \qquad < x > = \int \psi^* x \, \psi d\tau$$

In general, for any function $f(r)$, the expectation value

$$< f(r) > = \int \psi^* f(x) \, \psi d\tau$$

Since operator, $\text{E} \rightarrow i\hbar \frac{\partial}{\partial t}$

$$\therefore \qquad < \text{E} > = \int \psi^* i\hbar \frac{\partial}{\partial t} \psi d\tau$$

and operator, $p \rightarrow i\hbar \frac{\partial}{\partial t}$

$$< p_x > = \int \psi^* \left(-i\hbar \frac{\partial \psi}{\partial x} \right) d\tau$$

$$< p^2{}_x > = \int \psi^* \left(-\hbar^2 \frac{\partial^2 \psi}{\partial x^2} \right) d\tau$$

- ***Ehrenfest's theorem (Relationship between classical and quantum motion)***

$$\frac{d}{dt} < x > = \frac{1}{m} < p_x >$$

$$\frac{d}{dt} < p_x > = - < \frac{\partial V}{\partial x} >$$

- ***One dimensional wave packet : Concepts of group and phase velocities***

A wave packet obtained by the superposition of plane wave

$$\psi(x,t)=\frac{1}{\sqrt{2\pi}}\int_{-\infty}^{\infty} A(k)e^{i(kx-\omega t)}dk \qquad ...(1)$$

Now using $p = \hbar k$ and $E = \hbar\omega$. we get equation (1) as

$$\psi(x,t)=\frac{1}{\sqrt{2\pi\hbar}}\int_{-\infty}^{\infty} a(p)\exp\left[\frac{i}{\hbar}(px-Et)\right]dp$$

Thus

$$\psi(x,\ 0)=\frac{1}{\sqrt{2\pi\hbar}}\int_{-\infty}^{\infty} a\ (p)\ \exp\left(\frac{i}{\hbar}px\right)dp$$

$\frac{1}{\sqrt{2\pi\hbar}}$ has been introduced so that *a(p)* is given by symmetrical relation

$$a\ (p)=\frac{1}{\sqrt{2\pi\hbar}}\int_{-\infty}^{\infty} \psi(x,\ 0)\ \exp\left[-\frac{i}{\hbar}px\right]dx$$

- ***Group velocity***

$$v_g=\frac{d\omega}{dk}=\frac{dE}{dp}\left(U\sin g\ \begin{matrix}\bar{p}=\hbar k\\ \bar{E}=\hbar\omega\end{matrix}\right)$$

- ***Phase velocity***

$$v_p=\frac{\omega}{k}$$

The One Dimensional Gaussian Wave Packet : Propagation of Gaussian wave packet $\psi(x_1\ t = 0)$

$$=\frac{1}{\left(\pi\sigma^2\right)^{1/4}}\exp\left[-\frac{(x-x_0)^2}{2\sigma^2}\right]\exp\left[\frac{i}{\hbar}p_0x\right]$$

- Wave function is normalised
- $<x>=\int_{-\infty}^{\infty}\psi^* x\,\psi\,dx=x_0$
- $<x^2>\ =\int_{-\infty}^{\infty}\psi^* x^2\,\psi\,dx=\frac{\sigma^2}{2}+x_0^2$

- $<p> = \int_{-\infty}^{\infty} \psi^*\left(-i\hbar\frac{\partial\psi}{\partial x}\right)dx = p_0$

- $<p^2> = \int_{-\infty}^{\infty} \psi^*\left(-\hbar^2\frac{\partial^2}{\partial x^2}\psi\right)dx = \frac{\hbar^2}{2\sigma^2} + {p_0}^2$

Simple Solutions of the Schrodinger Equation

I : One Dimensional Consideration

Orthogonality Condition

$$\int_{-\infty}^{\infty} \psi_n{}^*\psi_n\, dx = 0$$

Normalization condition

$$\int_{-\infty}^{\infty} |\psi_n|^2\, dx = 1$$

Orthonormality condition

$$\int_{-\infty}^{\infty} \psi_k{}^*\psi_n\, dx = \delta_{kn}$$

$\delta_{kn} \rightarrow$ kronecker delta function

$$\delta_{kn} = 0,\ k \neq n = 1,\ k = n$$

Particle in a One-Dimensional Infinitely Deep Potential Well

The allowed energy levels are given by

$$E_n = \frac{\pi^2 n^2 \hbar^2}{2ma^2},\quad n = 1, 2, 3$$

$$= \frac{n^2 h^2}{8ma^2}$$

The corresponding eigenfunctions are

$$\psi_n = \sqrt{\frac{2}{a}}\sin\left(\frac{n\pi}{a}x\right)$$

Particle in a One-Dimensional Potential Well of Finite Depth

Energy eigenvalues would given as

$$E = \frac{(2p)^2 \pi^2 \hbar^2}{2ma^2} \; ; p = 1, 2, 3 \text{ for } V \to \infty$$

For $V_0 >> E$

$$E = \frac{(2p+1)^2 \pi^2 \hbar^2}{2ma^2}$$

Linear Harmonic Oscillator

The potential energy for this

$$V(x) = \frac{1}{2} m\omega^2 x^2$$

The energy eigenvalues

$$E = \left(n + \frac{1}{2}\right)\hbar\omega, \; n = 0, 1, 2, \ldots$$

Zero point energy

$$E = \frac{\hbar\omega}{2}$$

The normalised harmonic oscillator wave functions are

$$\psi_n(x) = N_n \exp\left(-\frac{1}{2}\alpha^2 x^2\right) H_n(\alpha x)$$

where, $$N_n = \left[\frac{\alpha}{\sqrt{\pi} 2^n \; n!}\right]^{1/2}$$

and $$\alpha = \left(\frac{m\omega}{\hbar}\right)^{1/2}$$

and $H_n(\alpha x) \to$ Hermite polynomial.

The Double Well Potential

For $E << V_0$

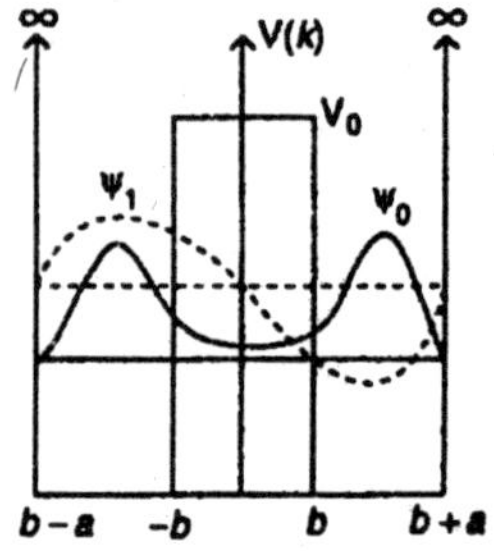

$$E_n^{(0)} = \frac{\pi^2\hbar^2 n^2}{2ma^2}$$

and $$E_n = E_n^{(0)} - \frac{2E_n^{(0)}}{k_0 a} \mp 4\frac{E_n^{(0)}}{k_0 a} e^{-2k_0 b}$$

where $$k_0 = \left[\frac{2m}{\hbar^2}\left(V_0 - E_n^{(0)}\right)\right]^{1/2}$$

For $V \to \infty,\ E_n \to E_n^{(0)}$

- ***Wave function***

$$\psi(x, t) = \frac{1}{\sqrt{2}} e^{-iE_n^{(1)}t/k} \left[\psi_S^{(1)}(x) + \psi_a^{(1)}(x) e^{-i\pi t/r}\right]$$

where

$\psi_s^{(1)} \to$ Lowest symmetric wave functions

$\psi_a^{(1)} \to$ Lowest antisymmetric wave functions

One Dimensional Barrier Transmission Problems

The Potential Step

- The current densities associated with the wave function Ae^{ikx}, Be^{-ikx} and ce^{ik_1x} represent the incident current, reflected current and transmitted current respectively, these are

$$J_{inc} = \frac{\hbar k}{m}|A|^2$$

$$J_{ref} = \frac{\hbar k}{m}|B|^2$$

$$J_{tr} = \frac{\hbar k_1}{m}|C|^2$$

$$\Rightarrow \qquad B = \frac{k - k_1}{k + k_2}A \text{ and } C = \frac{2k}{k + k_1}A$$

The reflection and transmission coefficients are

$$R = \frac{J_{ref}}{J_{inc}} = \frac{|B|^2}{|A|^2} = \frac{(k - k_1)^2}{(k + k_1)^2}$$

$$T = \frac{J_{tra}}{J_{inc}} = \frac{k_1|C|^2}{k|A|^2} = \frac{4kk_1}{(k + k_1)^2}$$

The Rectangular Potential Barrier

Transmission coefficient

$$T = \frac{-4p_1^2 p_2^2 \operatorname{sec}h^2(ip_2\, a\, \hbar)}{(p_1^2 + p_2^2)\tanh^2(ip_2\, a\, \hbar) - 4p_1^2 p_2^2}$$

$$R = \frac{(p_1^2 - p_2^2)^2 \tanh^2(ip_1\, a\, \hbar)}{(p_1^2 + p_2^2)^2 \tanh^2(ip_2\, a\, \hbar) - 4p_1^2 p_2^2}$$

A special case when the barrier is thick

$$T = \frac{-16p_1^2 p_2^2\, e^{2ip_2\, a/\hbar}}{(p_1^2 + p_2^2)^2 - 4p_1^2 p_2^2} = \frac{-16p_1^2 p_2^2\, e^{2ip_2\, a/\hbar}}{(p_1^2 - p_2^2)^2}$$

$$= \frac{+16E\,(V_0 - E)}{V_0^2} e^{\left[-2\sqrt{\{2m(V_0 - E)\}}a/\hbar\right]}$$

$$\mathrm{R} = \frac{(p_1^2 - p_2^2)^2}{(p_1^2 - p_2^2)^2} = 1$$

Simple Solutions of the Schrodinger Equation

II: 3-Dimensional Consideration

The hydrogen atom. The Hamiltonian for hydrogen atom

$$H = -\frac{\hbar^2}{2m_1}\nabla_1^2 - \frac{\hbar^2}{2m_2}\nabla_2^2 + V\left(\left|\overline{r_1} - \overline{r_2}\right|\right)$$

- The energy depends on the total quantum number n.
- For each n we have values of l ranging from 0 to $n-1$, and for each value of l, the m values range from $-l$ *to* $+l$ there

are $\sum_{l=0}^{n-1}(2l+1) = n^2$, states belonging to a particular energy.

For the Ground State of hydrogen atom the Wave Function is Spherically Symmetric

$$\psi_{100} = \frac{1}{\sqrt{4}}\left(\frac{z}{a_0}\right)^{3/2} \exp\left(-\frac{zr}{a_0}\right)$$

P(r)dr $\rightarrow$ the probability of finding the electron between r and $r + dr$ is

$$P(r)dr = |\psi_{100}|^2 4\pi r^2 dr$$

$$= 4\left(\frac{z}{a_0}\right)^3 r^2 \exp\left[-\frac{2zr}{a_0}\right] dr$$

The Particle in a Box (In 3-D)

- Wave function

$$\psi_{n_x n_y n_z} = \frac{2\sqrt{2}}{\sqrt{abc}} \sin\frac{n_x \pi x}{a} \sin\frac{n_y \pi y}{b} \sin\frac{n_z \pi z}{c}$$

- Energy levels

$$E_{n_x n_y n_z} = \frac{h^2}{8m}\left[\frac{n_x{}^2}{a^2} + \frac{n_y{}^2}{b^2} + \frac{n_z{}^2}{c^2}\right]$$

- Energy states between E and E + dE is

$$= 2\pi V\left(\frac{2m}{h^2}\right)^{3/2} E^{1/2}, \ V \rightarrow \text{Volume}$$

Three-Dimensional Harmonic Oscillator

For isotropic oscillator, energy is given as

$$E = \left(n+\frac{3}{2}\right)\hbar\omega$$

- The degree of degeneracy is $\frac{1}{2}(n + 1)\ (n + 2)$

Angular Momentum

- Angular momentum operator representations of L_x, L_y, and L_z and $L^2 = L_x{}^2 + L_y{}^2 + L_z{}^2$

$$L_x = -i\hbar\left(y\frac{\partial}{\partial z} - z\frac{\partial}{\partial y}\right)$$

$$L_y = -i\hbar\left(z\frac{\partial}{\partial x} - x\frac{\partial}{\partial z}\right)$$

$$Lz = -i\hbar\left(x\frac{\partial}{\partial y} - y\frac{\partial}{\partial x}\right)$$

- In terms of spherical polar coordinates

$$L_x = i\hbar\left(\sin\phi\frac{\partial}{\partial\theta} + \cot\theta\cos\phi\frac{\partial}{\partial\phi}\right)$$

$$L_y = i\hbar\left(-\cos\phi\frac{\partial}{\partial\theta} + \cot\theta\sin\phi\frac{\partial}{\partial\phi}\right)$$

and $$L_z = -i\hbar\frac{\partial}{\partial\theta}$$

$$L_y = -\hbar^2 r^2\left\{\nabla^2 - \frac{1}{r^2}\frac{\partial}{\partial r}\left(r^2\frac{\partial}{\partial r}\right)\right\}$$

Eigenvalues of L^2:

$$L^2 Y_{lm}(\theta, \phi) = l(l+1)\hbar^2 Y_{lm}(\theta, \phi)$$

eigenvalues of L^2 is $l(l + 1)\hbar^2$

and for $L_z \to m\hbar$

The Commutation Relations

- $$[L_x, L_y] = i\hbar L_z;$$
  $$[L_y, L_z] = i\hbar L_x;$$
  $$[L_z, L_x] = i\hbar L_y$$
- $[L^2, L_x] = [L^2, L_y] = [L^2, L_z] = 0$

  Angular momentum in general

  $$J = L + Se$$
  $$J^2 = J_x^2 + J_y^2 + J_z^2$$

  Ladder operators :

  $$J_+ = J_x + iJ_y$$
  $$J_- = J_x - iJ_y$$

  Then $[J_z, J_+] = \hbar J_+$

  $$[J_z, J_-] = -\hbar J_-$$

  and $[J_+, J_-] = 2\hbar J_z$

  $$[J^2, J_+] = [J^2, J_-] = 0$$

Expectation Value

$$< J^2 > = < J_x^2 > + < J_y^2 > + m^2\hbar^2$$

J_x and J_y are Hermitian and the expectation value of Hermitian are non-negative.

Operator	***Eigenvalue***	***Eigenket***
J^2	$\lambda\hbar^2 = j(j+1)\,\hbar^2$	$\|jm>$
J^2	$\lambda\hbar^2 = j(j+1)\,\hbar^2$	$J \pm \|jm>$
J_z	$m\hbar$	$\|Jm>$
	$(m+1)\hbar$	$J_+ \|jm>$
	$(m-1)\hbar$	$J_- \|jm>$

$\Rightarrow \quad J_+ \, |jm> \; = \; |j, m+1>$

and $\quad J_- \, |jm> \; = \; |j, m-1>$

Matrix Formulation

- Matrix element for J^2 and J_z are given as

$$< j'm' \left| J^2 \right| jm > \; = \; j(j+1)\,\hbar^2\, \delta_{jj'}\, \delta_{mm'}$$

$$< j'm' \left| J_z^2 \right| jm > \; = \; m\hbar\, \delta_{jj'}\, \delta_{mm'}$$

where $<j'm'$ |and| $jm>$ represents column and row respectively.

Thus each matrices of J^2 and J has the form

j	m \ j', m'	$j'=0$: $m'=0$	$j'=\frac{1}{2}$: $m'=\frac{1}{2}, -\frac{1}{2}$	$j'=1$: $m'=1, 0, -1$ ....
0	0	(element present)	(0)	(0)......
$\frac{1}{2}$	$\frac{1}{2}$	(0)	(element present)	(0)......
$\frac{1}{2}$	$-\frac{1}{2}$			
1	1			
1	0	(0)	(0)	(element present) ...
1	-1			

Note : J^2 and J_z matrices contain only principal diagonal

For Example :

1. $j = 1$

$J^2 =$

	j'	1	1	1
	m'	1	0	−1
j	m			
1	1	A	0	0
1	0	0	B	0
1	−1	0	0	C

[order of matrix = $(2j + 1)$
= $(2 \cdot 1 + 1)$
= 3]

$$A = B = C$$

$$= j(j+1)\hbar^2 = 1(1+1)\hbar^2 = 2\hbar^2$$

$$\therefore \qquad J^2 = \begin{pmatrix} 2\hbar^2 & 0 & 0 \\ 0 & 2\hbar^2 & 0 \\ 0 & 0 & 2\hbar^2 \end{pmatrix} = 2\hbar^2\begin{pmatrix} 1 & 0 & 0 \\ 0 & 1 & 0 \\ 0 & 0 & 1 \end{pmatrix}$$

and $\qquad J_z = m\hbar\,\delta_{jj'}\,\delta_{mm'}$ [As $m = 1, 0, -1$ for $j = 1$]

$$= \begin{pmatrix} 1.\hbar^2 & 0 & 0 \\ 0 & 0.\hbar^2 & 0 \\ 0 & 0 & -1.\hbar^2 \end{pmatrix}$$

$$= \hbar^2\begin{pmatrix} 1 & 0 & 0 \\ 0 & 0 & 0 \\ 0 & 0 & -1 \end{pmatrix}$$

2. For $\qquad j = \frac{1}{2}$ [order of matrix = 2 × 2]

$$J^2 = \frac{3}{4}\hbar^2\begin{pmatrix} 1 & 0 \\ 0 & 1 \end{pmatrix}$$

and $\qquad J_z = \frac{1}{2}\hbar\begin{pmatrix} 1 & 0 \\ 0 & -1 \end{pmatrix}$

- Matrix elements of J_x and J_y in a representation in which J^2 and J_Z are diagonal.

$$< j'm'|J_x|jm> = \frac{1}{2}\hbar\sqrt{\{(j-m)(j+m+1)\}}\,\delta_{jj'}\,\delta_{m',\,m+1}$$

$$\text{and} < j'm'|J_y|jm> = \frac{i\hbar}{2}\sqrt{\{(j+m)(j-m+1)\}}\,\delta_{jj'}\,\delta_{m',\,m-1}$$

- Matrix element for J_+, J_-

$$< j'm'|J_+|jm> = \hbar\sqrt{\{j(j+1)-m\,(m+1)\}}\,\delta_{jj'}\,\delta_{m',\,m+1}$$

$$\text{and} < j'm'|J_-|jm> = \hbar\sqrt{\{j(j+1)-m\,(m-1)\}}\,\delta_{jj'}\,\delta_{m',\,m-1}$$

For example

1. $$j = \frac{1}{2}$$

$$J_+ = \hbar\begin{pmatrix} 0 & 1 \\ 0 & 0 \end{pmatrix}$$

and $$J_- = \hbar\begin{pmatrix} 0 & 0 \\ 1 & 0 \end{pmatrix}$$

2. $$j = 1$$

$$J_+ = \sqrt{2}\,\hbar\begin{pmatrix} 0 & 1 & 0 \\ 0 & 0 & 1 \\ 0 & 0 & 0 \end{pmatrix}$$

and $$J_- = \sqrt{2}\,\hbar\begin{pmatrix} 0 & 0 & 0 \\ 0 & 0 & 0 \\ 0 & 1 & 0 \end{pmatrix}$$

- Matrix elements for L+ and L–

$$<l'm'|L_+|lm> = \hbar\sqrt{\{l(l+1)-m\,(m+1)\}}\,\delta_{ll'}\,\delta_{m',\,m+1}$$

$$< l'm'|L-|lm> = \hbar\sqrt{\{l(l+1)-m\,(m-1)\}}\,\delta_{ll'}\,\delta_{m',\,m-1}$$

For $j = 1$

l'	m'	m	1	0	−1
1	1	0	0	$\hbar\sqrt{2}$	0
1	0		0	0	$\hbar\sqrt{2}$
1	−1		0	0	0

$L_+ =$ (table above)

$$= \hbar\begin{pmatrix} 0 & \sqrt{2} & 0 \\ 0 & 0 & \sqrt{2} \\ 0 & 0 & 0 \end{pmatrix}$$

Similarly, $$L_- = \begin{pmatrix} 0 & 0 & 0 \\ \sqrt{2} & 0 & 0 \\ 0 & \sqrt{2} & 0 \end{pmatrix}$$

Spin Angular Momentum

- A particle having intrinsic (spin) angular momentum 'S' which is unrelated to its orbital motion. Also it obey

$$[S, L] = 0$$

- $$S^2 \,|SM_S> = s(s+1)\,\hbar^2\,|SM_S>$$

$$S_z\,|SM_S> = M_s\hbar^2\,|SM_S>$$

where $M_S = S, S-1, \ldots -s$, there are $(2s+1)$ values of M_S for given S and $(2s+1)$ states are possible

- For $s = \frac{1}{2}$; $M_s = \frac{1}{2}, -\frac{1}{2}$

Only $2\left(\frac{1}{2}\right)+1=2$ states are possible $|s, M_s>$

$\left|\frac{1}{2}, \frac{1}{2}>\uparrow\right.$ = α state $\left|\frac{1}{2}, -\frac{1}{2}>\downarrow\right.$ = β state

$$\therefore \quad \alpha = \begin{pmatrix} 1 \\ 0 \end{pmatrix} \uparrow \text{ and } \beta = \begin{pmatrix} 0 \\ 1 \end{pmatrix} \downarrow$$

α is wave function for $m_s = \frac{1}{2}$ or spin up while β stands for $m_s = -\frac{1}{2}$ or spin down wave function.

- General wave function

$$|x> \;=\; \begin{pmatrix} a \\ b \end{pmatrix} = a\begin{pmatrix} 1 \\ 0 \end{pmatrix} + b\begin{pmatrix} 0 \\ 1 \end{pmatrix} \;=\; a\alpha + b\beta$$

a →Probability amplitude that particle has spin up and

b →Probability amplitude for particle has spin down

- Also $S = \frac{1}{2}\hbar\sigma$

σ →Pauli matrices and defined as

$$\sigma_x = \begin{pmatrix} 0 & 1 \\ 1 & 0 \end{pmatrix};\ \sigma_y = \begin{pmatrix} 0 & -i \\ i & 0 \end{pmatrix};\ \sigma_z = \begin{pmatrix} 1 & 0 \\ 0 & -1 \end{pmatrix}$$

- $$S_+ = S_x + iS_y$$
$$S_- = S_x - iS_y$$

or,
$$S_+ = \frac{1}{2}\hbar\sigma_+ = \frac{\hbar}{2}\left[\sigma_x + i\sigma_y\right]$$
$$= \frac{\hbar}{2}\left[\begin{pmatrix} 0 & 1 \\ 1 & 0 \end{pmatrix} + \begin{pmatrix} 0 & 1 \\ -1 & 0 \end{pmatrix}\right]$$
$$= \frac{\hbar}{2}\begin{pmatrix} 0 & 2 \\ 0 & 0 \end{pmatrix} = \hbar\begin{pmatrix} 0 & 1 \\ 0 & 0 \end{pmatrix} \quad \text{Raising operator}$$

$$S_- = \frac{1}{2}\hbar\sigma_- = \hbar\begin{pmatrix} 0 & 0 \\ 1 & 0 \end{pmatrix} \quad \text{Lowering operator}$$

Properties of Pauli Spin Matrix

(1) $\sigma_x{}^2 = \sigma_y{}^2 = \sigma_z{}^2 = 1 \text{ and } \sigma^3 = 3$

(2) $\sigma_x\,\sigma_y = -\sigma_y\,\sigma_x = i\sigma_z$

(3) $\sigma_+{}^2 = \sigma_-{}^2 = 0$

(4) $\sigma_+ \sigma_- = 2(1+\sigma_z)$

(5) $(\bar{\sigma}.\bar{A})(\bar{\sigma}.\bar{B}) = (\bar{A}.\bar{B}) + i\,\bar{\sigma}.(\bar{A}\times\bar{B})$

Addition of Angular Momentum

- j_1 and j_2 are two independent angular momentum. The eigenvector $\psi\,(j_1 j_2 . m_1 m_2)$ defines eigenstates with four observable respectively $j_1\,(j_1+1)$, m_1, $j_2\,(j_2+1)$ and m_2.
- For given values of j_1 and j_2 *there* will be $(2j_1+1)$ possible values of m_1 and $(2j_2+1)$ possible values of m_2.

∴ There will be $(2j_1+1)\,(2j_2+1)$ eigenstates.

Clebsch-Gorden Coefficients (C-G Coeff.) Addition of Angular Momenta

$$\left.\begin{aligned}&\sum_{jm} < m_1 m_2 | JM >< JM | m_1' m_2' > \\ &\qquad = \delta_{m_1 m_1'}\,\delta_{m_2 m_2'} \\ &\sum_{m_1 m_2} < JM | m_1 m_2 >< m_1 m_2 | J' M' > \\ &\qquad = \delta_{JJ'}\,\delta_{MM'}\end{aligned}\right]\ \text{Definition}$$

Recurrence Relations for Calculation of C. G. Coefficient

(1) $$[J(J+1) - m(m+1)]^{1/2} < m_1 m_2 | JM+1 >$$
$$= [j_1(j_1+1) - m_1(m_1-1)]^{1/2} < m_1 - 1,$$
$$m_2 | JM > + [j_2(j_2+1) - m_2(m_2-1)]^{1/2}$$
$$< m_1,\ m_2 - 1 | JM >$$

(2) $$[J(J+1) - m(m-1)]^{1/2} < m_1 m_2 | Jm-1 >$$
$$= [j_1(j_1+1) - m_1(m_1+1)]^{1/2} < m_1 + 1,$$
$$m_2 | JM > + [j_2(j_2+1) - m_2(m_2+1)]^{1/2}$$
$$< m_2 + 1,\ m_1 | JM >$$

So to determine C.G. coefficient $<m_1 m_2 | JM>$ we use recurrence relation (1) and (2).

Operator Method in Quantum Mechanics

If the operator $\hat{\alpha}$, $\hat{\beta}$ and $\hat{\gamma}$ are of such that

$$\hat{\gamma} f(x) = \hat{\alpha} f(x) + \hat{\beta} f(x)$$

- The operator $(\hat{\alpha}\hat{\beta} - \hat{\beta}\hat{\alpha})$ is called the commutator of the operator $\hat{\alpha}$ and $\hat{\beta}$.

e.g., $$\hat{x}\,\hat{p}\,\psi(x) - \hat{p}\,\hat{x}\,\psi(x) = -\frac{h}{i}\psi(x)$$

$\therefore$ $$\hat{x}\,\hat{p} - \hat{x}\,\hat{p} = i\hbar$$

Theorem : The necessary and sufficient condition that simultaneous eigenfunctions exist for two operators $\hat{O}_1$ and $\hat{O}_2$ is that they commute.

- Momentum operator $\rightarrow -i\hbar \dfrac{d}{dx}$

  and a plane wave which represent a state of definite momentum p

$$\psi = \frac{1}{\sqrt{2\pi\hbar}} e^{(1/\hbar)\,p_x}$$

$\rightarrow$ Normalised eigenfunction

- If an operand u is a simultaneous eigenfunction of two operators $\hat{\alpha}$ and $\hat{\beta}$ such that

$$\hat{\alpha}\,u = \lambda u \text{ and } \hat{\beta} = \mu u$$

  where λ and μ are numerical constants, then u is also an eigenfunction of the operator $(\hat{\alpha}\,\hat{\beta} - \hat{\beta}\,\hat{\alpha})$ and belongs to the eigenvalue 0.

⇒ Operator associated with linear momentum

$$\bar{p} = -i\hbar\nabla$$

Energy operator

$$\bar{E} = i\hbar\frac{\partial}{\partial t}.$$

and velocity operator,

$$\bar{v} = -\frac{i\hbar}{m}\nabla$$

Hermitian Operator

For any two well behaved functions $f(x)$ and $g(x)$ vanishing at infinity, an operator $\hat{O}$ satisfying the equation

$$\int_{-\infty}^{\infty} g^*(x)\left[\hat{O}f(x)\right]dx = \int_{-\infty}^{+\infty}\left[\hat{O}g(x)\right]^* f(x)\,dx$$

is said to be Hermitian.

- ***Parity Operator*** π defined as

$$\hat{\pi}\psi(x) = \psi(-x)$$

Eigenvalue ⇒ ± 1

$+1 \rightarrow$ Corresponds to even eigenfunction.

$-1 \rightarrow$ Corresponds to odd eigenfunction.

⇒ $\hat{\pi}$ operator commutes with Hamiltonian operator $\hat{H}$

- ***Projection Operator***

$$\hat{p}i = |\psi_i><\psi_i| \rightarrow \text{Projection operator}$$

$$\sum_{i=1}^{\infty}\hat{p}_i = 1$$

Properties

(1) $$\hat{p}_i^2 = \hat{p}_i$$

(2) $$\hat{p}_i\hat{p}_j = \hat{p}_j\,\delta_{ij}$$

(3) Projection operator $\hat{p}_i$ is Hermitian

Unitary Operator : When the inverse and adjoint of an operator all identical, then the operator is known as unitary operator. If $\hat{U}$ is unitary operator, then

$$\hat{U}^+ = \hat{U}^{-1}$$

Properties of Hermitian Operations

(a) The sum of two Hermitian operators is Hermitian.

(b) The product of the Hermitian operators is Hermitian if and only if they commute.

(c) The operators as position $\hat{r}$, momentum

$\hat{p} = \left(\frac{\hbar}{i}\right)\nabla$ and energy $\hat{H} = \left(\frac{p^2}{2\mu}\right) + V(r)$ are Hermitian.

(d) Hermitian operators give real eigenvalues.

Theorem : The eigenfunctions of a Hermitian operator corresponding to different eigenvalues will be orthogonal function.

Equation of Motion

$$\dot{f} = [f, H]$$

or, $$\dot{x} = [x, H]$$

Theory of Scattering

The complete wave function in this case

$$\psi \underset{r\to\infty}{\longrightarrow} e^{i\underline{k}.\underline{r}} + f(\theta)\frac{e^{ikr}}{r}$$

$e^{ikr} \to$ incident wave and $f(\theta)\frac{e^{ikr}}{r}$ Scattered wave

where $f(\theta) \to$ Scattering amplitude

Differential scattering cross section

$$\sigma(\theta) = |f(\theta)|^2$$

Total scattering cross section (for spherically Symmetrical potential)

$$\sigma = 2\pi \int_0^{+\infty} \sigma(\theta) \sin\theta \, d\theta$$

The Born Approximation

Scattering amplitude

$$f(\theta, \phi) = -\frac{2\mu}{\hbar^2} \int V(r)\, e^{i\underline{k}.\underline{r}}\, d\underline{r}$$

If potential is spherically symmetric

$$f(\theta) = -\frac{1}{K} \int_0^{\infty} r \sin Kr\, V(r)\, dr$$

$$K = 2k \sin\frac{\theta}{2} \qquad k \to \text{propagation constant}$$

The differential scattering cross section

$$\sigma(\theta) = |f(\theta)|^2 = \left| -\frac{1}{K} \int_0^{\infty} r \sin Kr\, V(r)\, dr \right|^2$$

Condition for validity of Born approximation

$$\left| \frac{1}{4\pi} \int_0^\infty \frac{e^{ik|r-r'|}}{|\underline{r}-\underline{r}'|} \cup(r')\, e^{i\underline{k}.\underline{r}'}\, dr' \right| << 1$$

The Method of Partial Waves : The method of partial waves is mainly applicable to spherically symmetric potential and consists of the expansion of the wave function as a series of spherical harmonics multiplied by a radial wave function.

Scattering amplitude

$$f(\theta) = \sum_{l=0}^{\infty} \frac{(2l+1)}{k} \sin \delta_l\, e^{i\delta_l}\, P(\cos \theta)$$

The D.C.S. (Differential Scattering cross section)

$$\sigma(\theta) = \sum_{l=0}^{\infty} \frac{(2l+1)^2 P_l^2(\cos \theta) \sin^2 \delta_l}{k^2}$$

The T.C.S. (Total scattering cross section)

$$\sigma = \frac{4\pi}{k^2} \sum_{l=0}^{\infty} (2l+1)^2 \sin^2 \delta_l$$

Optical Theorem

$$\sigma_t = \frac{4\pi}{k}.\text{Im } f(0)$$

where $\quad \text{Im } f(0) = \frac{1}{k} \sum_{l=0}^{\infty} (2l+1)^2 \sin^2 \delta_l$

Scattering by Hard Sphere : For sufficiently low energies

TCS $\quad \sigma \approx 4\pi a^2$

For high energies

$$\sigma \approx 2\pi a^2$$

Perturbation Theory

Hamiltonian for He[$e^- - e^-$ interaction]

$$H = \sum_{i=1}^{2}\left(-\frac{\hbar}{2\mu}\nabla_i^{\,2},\ -\frac{ze^2}{r_i}\right)+\frac{e^2}{r_{12}}$$

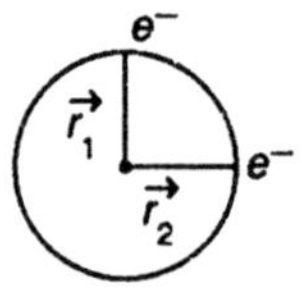

$\frac{e^2}{r_{12}}$ → Produces problem in solving the problem exactly.

This term is known as perturbation.

Time-Independent Perturbation Theory

- When a small disturbance is applied we can break Hamiltonian as

$$H = H^\circ + H'$$

H° → Non-perturbed, term

H' → Small perturbed term

(i) ***Non-degenerate Case :*** *(i.e.,* there will be one eigenfunction corresponding to each (energy level) eigenvalue)

In 1st Order Perturbation

1st order corrections are

(a) $$E = E_m + <m|H'|m>$$

$$<m|H'|m>$$

1st order correction in energy

(b) $$|\psi>^{(1)} = |m> + \underset{k}{S'}\frac{<k|H'|m>}{(E_m - E_k)}|k>$$

where $$|\psi_1> = \underset{k}{S'}\frac{<k|H'|m>}{(E_m - E_k)}|k>$$

1st order correction in wave function.

IInd Order Perturbation

$$W_2 = S'_k \frac{[< m|H'|k >]^2}{(E_m - E_k)}$$

This is IInd order correction in energy.

(ii) *Degenerate Case* : (Two or more eigenvalues corresponding to one eigenfunction).

1st order correction in energy

$$W_1 = \frac{(H'_{mm} + H'_{ll}) \pm \sqrt{(H'_{mm} - H'_{ll})^2 + 4H'_{ml}\,H'_{lm}}}{2}$$

Degeneracy is removed in 1st order.

where

$$H'_{mm} = < m|H'|m >$$

$$H'_{ll} = < l|H'|l >$$

$$H'_{ml} = < m|H'|l >$$

$$H'_{lm} = < l|H'|m >$$

1st Order Correction in Wave Function

$$|\psi_1 > = -S'_k \frac{[H'_{km}\,a_m + H'_{kl}\,a_l]}{(E_k - E_m)} |k >$$

IInd Order Correction in Energy

$$W_2 = \frac{1}{2} S'_n \frac{(H'^2_{mm} + H'^2_{ln})}{(E_m - E_n)}$$

$$\pm \frac{1}{2}\left[S'_n \left(\frac{H'^2_{mn} + H'_{ln}}{(E_m - E_n)} \right)^2 + \frac{4H'_{mn}\,H'_{ln}\,H'_{nm}\,H'_{nl}}{(E_m - E_n)^2} \right]^{1/2}$$

Degeneracy is removed in IInd order.

Zeeman Effect (Without e-Spin)

The effect of uniform external magnetic field on energy levels of an atom, is called the "Zeeman effect".

- Perturbation

$H' = \frac{e}{2\mu}(\vec{B} \cdot \vec{L})$ represents the energy $(-\vec{M} \cdot \vec{B})$ of a magnetic dipole with $\vec{M} = \frac{-e}{2\mu}\vec{L}$

So 1st order energy correction term

$$E_1 = \left< m \left| \frac{e}{2\mu} BL_z \right| m \right> = \frac{eB}{2\mu} m\hbar$$

So, for a given n, m can take value from $-l$ to l *i.e.*, the degeneracy is of $(2l + 1)$ states. So, energy levels are splited by applying magnetic field. The nth level being splited into $(2l + 1)$ levels. This is Zeeman effect.

Stark Effect : The splitting of energy levels of an atom caused by a uniform external electric field $\vec{E}$ is called Stark effect.

- For ground state there is no 1st order Stark effect.
- For 1st excited state, $n = 2$

There is four fold degeneracy. For which we have four roots 0, 0, $3e\,Ea_0$ and $-3eEa_0$. So, two roots 0, 0 are same and hence two fold degeneracy have been removed by 1st order Stark effect.

So, behaviour of hydrogen atom in 1st excited state $(n - 2)$ is like a permanent electric dipole moment of, magnitude $3ea_0$ that can be mapped in three different ways.

(1) One state parallel to external field.

(2) One state antiparallel to field.

(3) Two states with zero component along field.

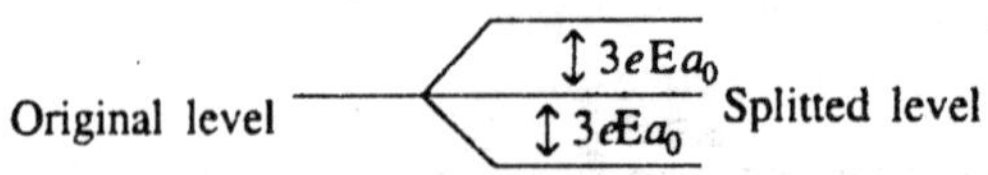

Golden Rule of Fermi

Transition probability per unit time is

$$T = \frac{2\pi}{\hbar}\left|H_{ml}^{(1)}\right|^2 \rho(m)$$

$$T = \frac{2\pi}{\hbar}\left|H_{ml}^{(1)}\right|^2 \rho(m)$$

$$H_{ml}^{(1)} = < m\left|H^{(1)}\right| l >$$

This is Fermi's Golden rule and states that the transition probability per unit time

(i) Is non-zero only between continuum states of the same energy.

(ii) Is proportional to the square of $\left|H_{ml}^{(1)}\right|$ of the perturbation connecting the states.

(iii) Is proportional to the density of final states.

- The probability of finding the system in *m*th state—

Considering the *m*th level above *l*th level, then

$\omega_{ml} = (E_n - E_l)/\hbar$ = positive quantity

If *l*th level above *m*th level, then

$\omega_{ml} = (E_n - E_l)/\hbar$ = negative quantity

So, when $E_m > E_l$, the probability is

$$= \frac{4\left|H^{\circ(1)}_{ml}\right|^2}{\hbar^2} \cdot \left[\frac{\sin^2(\omega_{ml} - \omega)t_0/2}{(\omega_{ml} - \omega)^2}\right]$$

When $El > E_m$, then the probability is

$$= \frac{4\left|H^{\circ(1)}_{ml}\right|^2}{\hbar^2}\left[\frac{\sin^2(\omega_{ml} - \omega)t_0/2}{(\omega_{ml} - \omega)^2}\right]$$

Identical Particles and Spin

Postulates

(1) The identical particles having an integral quantum number for their intrinsic spins are described by the symmetric wave function of type

$$\psi(1, 2, 3, \ldots, r, \ldots s, \ldots, N) = +\psi(1, 2, 3, \ldots s, \ldots r, \ldots N)$$

The particles described by this type of wave function are known as Bose particles (Bosons). Examples of Bosons are photon (spin 1), neutral helium atom in their normal state (spin 0) and α-particle (spin 0) etc.

(2) The particles with half odd integral quantum number for their intrinsic spins are described by antisymmetric wave function and given as

$$\psi(1, 2, 3, \ldots, r, \ldots s, \ldots, N) = -\psi(1, 2, 3, \ldots s, \ldots r, \ldots N)$$

This type of particle are known as Fermi particle (Fermions) because they obey Fermi-Dirac statistics.

Examples : Electrons, protons, neutrons, muons (all with spin 1/2)

- Antisymmetric wave function can be built by adding all the functions with even number of interchanges and subtracting the sum of all those with an odd number of interchanges.

Pauli's Exclusion Principle

- The probability of two antisymmetric identical particles is zero or two non-interacting Fermions cannot both be in same quantum state.

2

Elementary Particles

Cosmic Rays

Cosmic rays are highly penetrating radiations consisting of high-energy atomic nuclei which are continuously coming from outer space. About 10^{18} of them reach the earth each second. They have a wide range of energy from 10^9 to 10^{18} electron-volts.

Elster and Geital in 1899 and Wilson in 1900 first detected the cosmic rays. They found that the charge on a very well-insulated electroscope always 'leaked' away in time although there was no ionising agent nearby which could make the air conducting. It was then thought that the leakage of charge was due to the ionising radiation from radioactive minerals in the earth.

Hess, in 1911, sent electroscope up in ballon and found that the intensity of the unknown ionising radiation was larger at high altitudes than at the surface of the earth. He therefore proposed that the radiation originated from somewhere outside the earth's atmosphere. The radiation was finally called 'cosmic rays' by Millikan in 1925.

From 1926 to 1928, Millikan and Cameron conducted experiments on the penetration of cosmic rays through water by lowering the electroscope in deep lakes. They found that the cosmic rays reaching the earth consisted of two components, soft and hard, of very different penetrating powers.

The soft component (now known to consist of electrons, positrons and photons) was absorbed at a depth of about 1 meter of water, but the more penetrating hard component (now known to consist of mesons) was not fully absorbed even at the bottom of the 500-meter deep lake. This showed that the energy of the cosmic rays was higher than any radiation known at that time.

Primary and Secondary Cosmic Rays : Investigations have shown that there are two types of cosmic rays : primary cosmic rays and secondary cosmic rays.

The primary cosmic rays are those which come from outer space and enter the upper part of the earth's atmosphere. They consist of atomic nuclear which have lost their orbital electrons. Most of them are hydrogen nuclei (protons), some are helium nuclei (α-particles), and a few are the nuclei of heavier elements. Their energies are distributed over a wide range.

The secondary cosmic rays are produced in the upper atmosphere by the collision of primary rays with the nuclei of the atmospheric gases. Thus *the rays found near earth's surface are the secondary cosmic rays.* These rays consist chiefly of μ-mesons, and positive and negative electrons. When a primary cosmic-ray proton collides with an air nucleus in the upper atmosphere, it produces mostly π-mesons (positive, negative and neutral) and some hyperons.

The air nucleus is left in an excited state from which the remaining nucleons (protons and neutrons) come out. These secondary nucleons and charged π-mesons multiply further by undergoing similar nuclear collisions.

Many of the charged π-mesons decay into μ-mesons :

$$\pi^{\pm} \rightarrow \mu^{\pm} + v \qquad \text{(neutrino).}$$

The μ-mesons do not interact with nuclei and arrive at sea level with very little energy loss. Many of them even penetrate deep into the ground. Hence *at sea level, about 80% of the (secondary) cosmic rays are μ-mesons.*

Cosmic-ray Showers : The neutral π-mesons produced in the upper atmosphere decay into γ-ray photons :

$$\pi^{o} \rightarrow \gamma + \gamma.$$

These γ-rays in turn produce pairs of positive and negative electrons :

$$\gamma \rightarrow e^{+} + e^{-}.$$

The electrons radiate more photons which, in turn, produce pairs. Thus a cascade shower process continues, whose alternate links are electron pair and photon. Such showers were actually seen by Blackett and Occhialini in 1933 in cloud-chamber photographs.

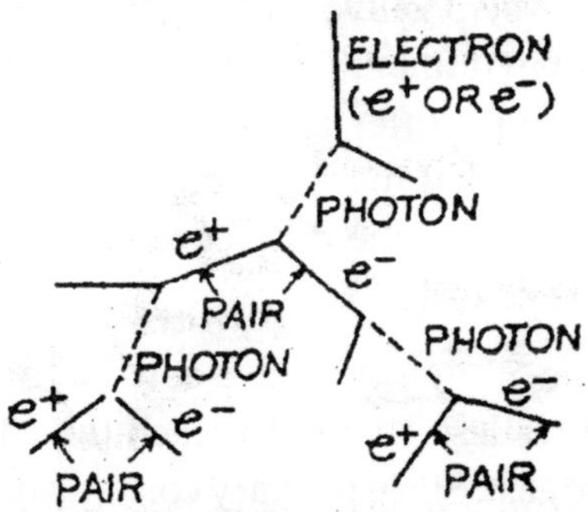

Variations in Cosmic Rays : The intensity of cosmic rays varies with latitude, direction and altitude.

Latitude Variation

The intensity of cosmic rays is greatest and practically uniform in the regions of earth's magnetic poles ; it decreases more and more rapidly as the equator is approached.

This variation is easily explained. As the primary cosmic rays, which are positively-charged particles, approach the earth's atmosphere vertically, their paths are deflected away from the normal by the earth's magnetic field. This deflection varies with magnetic latitude. Particles falling along the magnetic axis of the earth (towards the magnetic poles) do not suffer any deviation and reach the earth even if they have very small energy.

Therefore, the number of cosmic-ray particles reaching the magnetic poles is maximum. Particles falling at an angle with the earths magnetic axis describe a helical path, and those moving very slowly may be bent so much that they do not reach the earth's surface. Those arriving on the magnetic equator suffer the largest deflection because they are moving perpendicular to the magnetic field. So, only most energetic particles can reach the earth at the equator.

Directional Variation : East-West Asymmetry : Particles of opposite signs are bent in opposite directions by earth's magnetic field. If the positive and negative particles in cosmic rays are different in number, then the cosmic rays reaching the earth from west will differ in intensity from those reaching from east. It is seen that more cosmic ray particles come from the west than from east, the difference being maximum at the equator.

Calculations show that a positive particle coming from west requires less energy to reach the equator than that coming from the east. Since primary cosmic rays are positively-charged, they are expected more frequently from west than from east.

Altitude Variation

The intensity of cosmic rays increases with altitude. It, however, reaches a maximum not at the top of the atmosphere but well below it, above which the intensity again drops.

The position of the maximum indicates that much of the cosmic radiation must actually be produced there. This happens as the primary rays suffer nuclear collisions, producing secondary rays.

Origin of Cosmic Rays

The origin of primary cosmic rays is still an unsolved problem. We do not know definitely where cosmic rays come from and how they acquire their tremendous energies.

One of the views is that some of the cosmic rays may come from the sun. According to this view, some of the protons in the sun acquire high energies at times of solar activity when great eruptions take place and masses of ionised gases shoot out from the sun into interplanetary space. The main argument against this assumption is the uniform intensity of cosmic rays at all hours of the day and night. Counter-arguments are, however, also in progress.

Another view is that the cosmic rays come from the stars. This view is based on the fact that the total energy flux at the top of the atmosphere due to cosmic rays is about equal to the radiant energy flux received from all stars excluding the sun. This view is also unacceptable because every star cannot emit equally the light and the cosmic rays.

It has also been suggested that some special class of stars known as 'flare stars', 'magnetic variables' and 'supernova' may be responsible for most of the cosmic rays.

Most scientists believe, however, that cosmic-ray particles are accelerated by electro-magnetic processes as take place in solar flares. Stars have been observed in which eruptions resembling solar flares occur. Double stars with large magnetic moments and revolving around a common centre have also been observed. All of these are possible sources of cosmic rays.

Elementary Particles : *An elementary particle is one whose internal structure cannot be described as a combination of other particles.*

Until 1932, the only elementary particles known were the positively-charged proton (p), the negatively-charged electron (e) and the photon (γ). In that year two new elementary particles were discovered, namely, the neutron (n) and the positive electron or positron (e^+).

From 1932 onwards the number of elementary particles known has increased continuously. The negatively and positively charged mu-mesons or muons (μ^- and μ^+) were discovered in 1936, charged pi-mesons or pions (π^+ and π^-) in 1947, a neutral pion (π°) in 1950 and the negatively-charged counterpart of the proton known as antiproton $(\overline{p})$ in 1955. Then came the antineutron $(\overline{n})$, a new family of charged and neutral mesons known as *K*-mesons or kaons (K^+, K^-, K°), and hyperons (particles heavier than proton). The hyperon family includes lambda-hyperon ($\wedge^\circ$), sigma-hyperons (Σ^+, Σ^-, Σ°), xi-hyperons (Ξ^-, Ξ°) and omega-hyperon (Ω°).

The existence of a neutral, massless particle neutrino (v) was suggested by Pauli in 1930 and theoretically established by Fermi in 1934. The particle was actually discovered in 1956. Further in 1963, it became known that there are two kinds of neutrinos, the electron-neutrino (v_e) and the muon-neutrino ($v\mu$).

Characteristic Properties of Elementary Particles : The elementary particles are characterised by their mass, charge, life-time, spin, etc.

Mass : An elementary particle has always the same rest mass. If two particles have different rest masses, we consider them different particles. The magnitude of the rest mass serves as the principal label which identifies the particle uniquely. The value of the rest mass may allow us to infer the existence of a particle even without a direct observation, just from the conservation of energy and momentum. The neutral pion π° was discovered in this very way.

Charge : All known elementary particles have charge $+e, -e$, or zero. Further, the charge is always conserved in any collision process. This is clear in the following neutron-proton collisions :

$$n + p \rightarrow p + p + \pi^-.$$

$$n + p \rightarrow n + n + \pi^+$$

$$n + p \rightarrow n + p + K^- + K^+.$$

Average Lifetime : All elementary particles, except photon, electron, proton and neutrinos are unstable and undergo decay into other elementary particles of smaller mass. The decay probability of a particular particle is, however, independent of the length of line the particle has lived. It is impossible to predict when a given (unstable) particle will decay. Hence an elementary (unstable) particle has an *average* life-time, which is independent of the way the particle decays.

Spin : Many elementary particles spin in a manner analogous to that of the earth on its axis, but with certain differences. The spin property forms a basis for the classification of elementary particles.

Interactions : Four kinds of interactions between elementary particles are known : *gravitational, weak, electromagnetic and strong*. Their relative magnitudes are in the ratios 10^{-39}; 10^{-13}; 10^{-3}: 1. The (weakest) gravitational interaction is universal. All elementary particles without any exception are subjected to it.

It is the interaction that holds the earth together, binds the sun and planets into the solar system, and binds stars into galaxies. It is, however, insignificant in dealing with atoms and nuclei.

The weak interaction causes the light particles to interact with one another and with heavier particles. It is responsible for particle decays in which neutrinos are involved, notably beta decays. The electromagnetic interaction binds electrons to atoms and binds atoms together to form molecules and crystals.

The strong interaction is found between heavier particles. For example, it binds neutrons and protons to form the nuclei of all the elements. It is the dominant interaction in high-energy particle physics.

Antiparticles : The Dirac's relativistic quantum theory of electron predicted the existence of what is called an antiparticle. The antiparticle of a given particle has exactly the same mass, spin and life-time (if unstable) but *opposite* charge (if any). Further, the alignment between the spin and magnetic moment of an antiparticle is *opposite* to that of the corresponding particle. Further, when an antiparticle meets its particle, they mutually annihilate.

The first antiparticle known was the positron which was discovered in 1932 in a cloud chamber exposed to cosmic rays. It is a particle having the same mass as an electron, but opposite charge. When a positron comes to rest in matter, it is quickly annihilated by an electron resulting in two γ-photons :

$$e^+ + e^- \rightarrow 2\gamma.$$

The γ-photons fly away in different directions such as to conserve both momentum and energy.

Conversely, when a γ-photon of energy greater than 1.02 MeV passes near a nucleus in a substance, an electron and a positron are created (pair production):

$$\gamma + X \rightarrow e^- + e^+ + X,$$

where X is the nucleus in whose force field the pair is born. Any available energy in excess of 1.02 MeV goes into the kinetic energies of the electron and the positron. The presence of a nucleus X is required in order that momentum as well as energy be conserved in the process. The nucleus receives the excess momentum of the γ-photon over the total momentum of an electron and a positron.

The antiparticle of a proton is called the antiproton (or negative proton) and is denoted by $\overline{p}$. It was first produced in 1955 in proton-proton and proton-neutron collisions :

$$p + p \rightarrow p + p + p + \overline{p}$$

$$p + n \rightarrow p + n + p + \overline{p}$$

The energy of the bombarding proton on metal targets was of the order of 6 GeV (1 GeV = 10^9 eV). An antiproton differs from a proton in the sign of its electric charge and in its magnetic moment (in an antiproton the magnetic moment is directed oppositely to the mechanical angular momentum).

Antineutrons ($\overline{n}$) were observed in 1956 in certain antiproton-proton collisions :

$$p + \overline{p} \rightarrow n + \overline{n} \, .$$

An antineutron differs from a neutron in the sign of its instrinsic magnetic moment (in an antineutron the direction of the magnetic moment coincides with that of the mechanical angular momentum).

The annihilation of a particle-antiparticle pair need not always result in a pair of γ-photons (energy), as it does in the case of electron-positron annihilation. When an antiproton is annihilated with a proton, for example, several neutral and charged re-mesons are usually produced :

$$p + \overline{p} \rightarrow \pi^+ + \pi^- + \pi^+ + \pi^- + \pi^o$$

$$p + \overline{p} \rightarrow \pi^+ + \pi^- + \pi^o + \pi^o + \pi^o$$

An antiproton may annihilate when it encounters not only a proton, but also a neutron. For example, the following process is possible :

$$\overline{p} + n \rightarrow \pi^+ + \pi^- + \pi^- + \pi^o + \pi^o$$

Similar annihilations are observed for neutrons.

The distinction between the neutrino and the antineutrino is particularly interesting. The neutrino spins counter clockwise when viewed from behind, while antineutrino spins clockwise.

There are particles that are identical with their antiparticle (*i.e.* they have no antiparticles). Such particles are called 'absolutely neutral'. These include the photon, π^o-meson and the η-meson. These particles are not capable of annihilation. All the other particles do have their antiparticles.

Classification of Elementary Particles : Elementary particles are usually classified into four groups according to their mass and spin. These are :

1. *Photon* (γ) : Photon forms a group by itself. It is a stable particle with zero charge and zero rest mass. It has a spin equal to 1, and is therefore a boson. It is the quantum of the electromagnetic field and participate in electromagnetic interactions.
2. *Leptons* : These are light-weight particles. They include electrons (e^-, e^+), muons (μ^-, μ^+), electron-neutrinos (v_e, $\overline{v}_e$) and muon-neutrinos (v_μ, $\overline{v}_\mu$). All leptons have a spin equal to ½, and are therefore fermions. They interact weakly with other particles. Except muons, the leptons are stable.

3. *Mesons* : These are middle-weight particles having masses intermediate between the muons and the nucleons (protons and

Class	Name	Particle symol	Anti-particle	Spin	Restmass (m_e)	average-life (sec)
Photon	Photon	γ	(γ)	1	0	stable
	Eloectron	e	e	1/2	1	stable
Leptons	μ-meson	μ	μ	1/2	207	2.2 × 10[illegible]
	e-neutrino	v_e		1/2	0	stable
	μ-neutrino	v_μ		1/2	0	stable
HADRONS	Positive π-meson	π^+	π	0	273	2.6× 10[illegible]
	Neutral π-meson	π^o	(π^o)	0	264	9× 10[illegible]
Mesons	Positive K-meson	K	K^-	0	966	1.2 × 10[illegible]
	Neutral K-meson	K^o		0	974	8.6 × 10[illegible] & 5.2 × 10[illegible]
	η–meson	η''	(η'')	0	1073	2.4 × 10[illegible]
	Nucleons { Proton	p		1/2	1836	stable
	Nucleons { Neutron	n		1/2	1839	930
	Lambda hyperon	$\wedge^o$		1/2	2182	2.5 × 10[illegible]
	Positive sigma--hyperon	Σ^+		1/2	2328	8.0 × 10[illegible]
Baryons	Neutral " "			1/2	2332	< 10[illegible]
	Negative " "	Σ		1/2	2341	1.5 × 10[illegible]
	Negative xi hyperon	Ξ		1/2	2571	1.7 × 10[illegible]
	Neutral " "	Ξ^o		1/2	2583	3.0 × 10[illegible]
	Negative omega--hyperon	Ω^-		3/2	3290	1.3 × 10[illegible]

neutrons). They include pions (π^+, π^-, π^o), kaons (K^+, K^-, K^o, $\overline{K}^o$), and the eta-meson (η^o). The spin of all mesons is zero, so that they are bosons. All mesons are unstable.

4. *Baryons* : These are the heavy-weight particles which include nucleons (p, $\overline{p}$, n, $\overline{n}$) and hyperons ($\wedge^0$, Σ^+, Σ, Σ^o, Ξ^-, Ξ^o, Ω^-). The spin of all baryons is ½ except that of the Ω hyperon, which is $\frac{3}{2}$. Thus baryons are fermions. Except for the proton, all baryons are unstable. When a baryon decays, another baryon is formed without fail in addition to other particles. This is one of the manifestations of the law of baryon charge conservation.

Mesons and baryons are strongly-interacting particles, and collectively they are called 'hadrons'.

The classifying table shows particles as well as antiparticles. When the antiparticle is designated with the aid of a bar (—), the name of the antiparticle is obtained by adding the prefix "anti" to the name of the relevant particle. For example, the antiparticle of the lambda hyperon is called antilambda hyperon. The antiparticle of the electron is the positron. In the remaining cases, the names of particles and antiparticles are distinguished by adding the words 'positive' and 'negative'. For example, the negative pi-meson is the antiparticle of the positive pi-meson. Those antiparticles which are identical with their particles are shown in brackets.

Besides above, there are elementary particles predicted theoretically but not yet discovered, e.g., quarks which have fractional charges.

Leptons : These are light-weight elementary particles which are stable, or decay only by weak interactions. They are all fermions (having an intrinsic spin of ½) and occur as particles and antiparticles. Leptons include electrons, μ-mesons, electron-neutrinos and muon-neutrinos and their antiparticles.

(i) *Electrons and Positrons :* Electron is a stable atomic particle of mass of 9.1×10^{-31} kg and negative charge of 1.6×10^{-19} coulomb. Its antiparticle positron is identical with electron except that it is positively charged. When electron and positron meet, they annihilate each other, producing two γ-photons :

$$e^+ + e^- \rightarrow \gamma + \gamma.$$

(ii) *Ma-mesons (or Muons) :* Mu-mesons, discovered in 1936 by Anderson and Neddermeyer, are the components of cosmic radiation which are found in abundance in the cosmic rays at the ground level. They are created as a result of pi-meson decay in cosmic radiation.

Mu-mesons exist as both negative and positive and are denoted as μ^- and μ^+ respectively. Both have the same rest mass of 207 m_c. (intermediate between the rest mass of an electron and the rest mass of a proton) and the same spin of ½. The μ^+ meson is the antiparticle of μ^- meson. There is no neutral mu-meson.

Both μ^- and μ^+ mesons are unstable, having an average life of 2.2×10^{-6} sec. They decay spontaneously into electron (or positron) according to the following scheme :

$$\mu^- \rightarrow e^- + \overline{v}_e + v_\mu$$

$$\mu^+ \rightarrow e^+ + v_e + \overline{v}_\mu .$$

The neutrino and antineutrino are not actually observed, their emission is postulated to conserve both energy and momentum. An energy of 105 MeV is released in this decay.

Unlike pi-mesons (which interact strongly with nuclei), the μ-mesons interact *weakly* with nuclei. Hence they penetrate considerably through matter before being absorbed. Generated high in the atmosphere, most of them reach sea-level and some penetrate considerable distances into the earth. Hence the hard component of secondary cosmic rays consist mostly of μ-mesons.

The negative mu-meson can replace electron in atomic Bohr orbit to form 'mesic' (or 'muonic') atom. Thus, apart from its extra mass, a mu-meson behaves like an electron.

(iii) *Electron-neutrinos and Muon-neutrinos* : These particles, denoted by v_e and v_μ , have negligible rest mass and no charge, and are spin ½ fermions. They are distinguished from their anti-particles $\overline{v}_e$, and $\overline{v}_\mu$ by their *helicities* which are left-handed for neutrinos and right-handed for antineutrinos. They participate in weak interaction with matter.

Mesons : These are middle-weight particles having masses intermediate between the muons and the nucleons (protons and neutrons). They are all bosons having zero spin and are unstable. They include pions (π^+, π^-, π°), kaons ($K^+, K^-, K^\circ, \overline{K}^\circ$) and the eta-meson ($\eta^\circ$).

(i) *Pi-mesons (or Pions) :* Pi-mesons were discovered in 1947 in the upper atmosphere, although their existence was predicted much earlier by Yukawa in his meson theory of nuclear forces. They are created in nature when primary cosmic-ray protons collide with atmospheric nuclei and cause energetic nuclear disintegrations. After the development of high-energy accelerators, pi-mesons have also been created artificially.

Pi-mesons can exist in all three states; π^+, α^- and π°, all having a zero spin. The π^+ and π^- mesons have rest masses of 273 m_e, while that of the π° mesons is slightly less, 264 m_e. The π^- meson is the antiparticle of the π^+ meson, while the π° meson is its own antiparticle.

Pions interact strongly with nucleons. For example, they are produced by collisions of high-energy (several hundred MeV) protons produced in synchrocyclotron. Some pion production reactions are

$$p + p \rightarrow p + n + \pi^+$$

$$p + n \rightarrow p + p + \pi^-$$

$$p + \gamma \rightarrow n + \pi^+$$

$$p + \gamma \rightarrow p + \pi^o$$

The pi-mesons are highly unstable. The average life-time of (charged) π^+ and π^- mesons is only 2.6×10^{-8} sec, while that of (neutral) π° meson is still shorter, 9×10^{-17} sec. These life-times are so short that only a fraction of cosmic ray pi-mesons can reach earth. Charged pi-mesons decay in flight by weak interaction into lighter mu-mesons and mu-neutrinos :

$$\pi^{+} \rightarrow \mu + v_{\mu}$$

$$\pi^{-} \rightarrow \mu- + \bar{v}_{\mu}$$

These mu-neutrinos are somewhat different from the neutrinos involved in beta-decay which are better called electron-neutrinos. The emission of the mu-neutrinos is postulated to conserve both energy and momentum.

As the rest mass of a mu-meson is 207 m_e and that of neutrino is zero, a mass-energy of $(273 - 207)m_e = 66\ m_e$ is available in the above decay. As m_e is equivalent to 0.51 MeV, the energy released is $66 \times 0.51 = 33.7$ MeV. Of this energy, a constant amount of 4 MeV appears as the kinetic energy of the mu-meson. The mu-mesons are also unstable and decay into electrons.

The neutral pi-meson decays spontaneously by an electro-magnetic interaction into two high-energy γ photons :

$$\pi^{\circ} \rightarrow \gamma + \gamma.$$

The energy released is shared equally between the two γ-photons.

(ii) *K-mesons (or Kaons)* : There is also a heavier class of mesons, called K-mesons, which were first discovered in high-altitude cosmic ray experiments but now are available from the accelerators. They exist as K^+ and its antiparticle K^-, and also as K° and its antiparticle $\overline{K}^{\circ}$. The charged K-mesons (K^+ and K^-) have rest masses of 966 m_e, spin of zero, and mean-lives of 1.2×10^{-8} sec.

K-mesons are easily produced through strong interaction. An example is the reaction

$$\pi^{-} + p \rightarrow \Lambda^{\circ} + K^{\circ}$$

The charged meson decay in a variety of ways :

$$K^+ \rightarrow \pi^+ + \pi^+ + \pi^-$$

$$K^\pm \rightarrow \pi^\pm + \pi^o$$

$$K^\pm \rightarrow \mu^\pm + v_\mu$$

There are two distinct varieties of neutral K-mesons, $K_1°$ and $K_2°$. Both have rest masses 974 m_e, but half-lives of the order of 10^{-10} and 10^{-8} sec respectively. They can decay in several modes :

$$K_1° \rightarrow \pi^+ + \pi^-$$

$$\rightarrow \pi^o + \pi^o$$

$$K_2° \rightarrow \pi^+ + \pi^- + \pi^o$$

$$\rightarrow \pi^o + \pi^o + \pi^o.$$

(iii) *η-mesons :* The neutral eta-meson η° was discovered in 1961. It has a rest mass of 1073 m_e, which is intermediate between that of a K-meson and that of a nucleon. Like others, the η-meson is also a zero spin boson. It decays electromagnetically in about 10^{-19} sec into two photons.

Moons and Pions

Muons and pions are elementary particles having rest masses intermediate between the rest mass of an electron and that of a nucleon (proton or neutron). The muon was discovered in 1936 in the cosmic rays at ground level, while the pion was discovered in 1947 also in cosmic rays but in the upper atmosphere although its existence had been predicted by Yukawa in 1935 in connection with nuclear forces.

Although the rest masses of muon and pion are close to each other, but they so much differ in their spin and interaction properties that they have been placed in different classes of elementary particles. The muon is a lepton, while the pion is a meson. Their comparative properties are as under:

	MUON	PION
1. Kinds:	μ^-, μ^+	π^+, π^-, π^o
2. Rest Mass :	207 m_e	273 m_e (π^+, π^-); 264 m_e (π^o)
3. Spin:	½ (fermion)	0 (boson)
4. Interaction with nuclei:	weak	strong
5. Half-life (sec):	2.2×10^{-6}	2.6×10^{-8} (π^+, π^-); 9×10^{-17} (π^o)
6. Decay mode :		
	$\mu^- \rightarrow e^- + \overline{v}_e + v_\mu$	$\pi^+ \rightarrow \mu^+ + v_\mu$
	$\mu^+ \rightarrow e^+ + v_e + v_\mu$	$\pi^- \rightarrow \mu^- + \overline{v}_\mu$
		$\pi^o \rightarrow \gamma + \gamma$

Baryons : These are heavy-weight elementary particles which include nucleons (protons and neutrons) and particles heavier than nucleons. They are all fermions and are strongly interacting. Except proton, all baryons are unstable. The two sub-classes of baryons are :

(i) *Nucleons :* These are the nuclear particles proton *(p)*, neutron (*n*) and their antiparticles, antiproton ($\overline{p}$) and antineutron ($\overline{n}$). The mass of proton is 1836 m_e and that of neutron is 1839 m_e. They all have a spin of ½.

(ii) *Hyperons :* Elementary particles heavier than nucleons are called 'hyperons'. They are unstable and have an average life-time of the order of 10^{-10} sec. The known hyperons fall into four classes :

(1) lambda hyperons Λ^o (with zero charge) and their antiparticles.

(2) sigma hyperons Σ^+, Σ^o, Σ^-, (with positive, zero and negative charges) and their antiparticles.

(3) xi hyperons Ξ^-, Ξ^o (with negative and zero charges) and their antiparticles.

(4) Omega hyperons Ω^- (with negative charge) and its antiparticle.

The spin of all hyperons is ½ except that of Ω-hyperon, which is $\frac{3}{2}$. Some possible decay modes for the hyperons are as follows :

$$\begin{aligned}
\Lambda^o &\rightarrow p + \pi^- \\
&\rightarrow n + \pi^o \\
\Sigma^+ &\rightarrow p + \pi^o \\
&\rightarrow n + \pi^+ \\
\Sigma^o &\rightarrow \Lambda^o + \gamma \\
\Sigma^- &\rightarrow n + \pi^- \\
\Xi^- &\rightarrow \Lambda^o + \pi^- \\
\Xi^o &\rightarrow \Lambda^o + \pi^o \\
\Omega^- &\rightarrow \Xi^o + \pi^- \\
&\rightarrow \Xi^- + \pi^o \\
&\rightarrow \Lambda^o + K^-
\end{aligned}$$

Conservation Laws Governing Elementary Particles : The modes of decay of the elementary particles are governed by the conservation laws of *charge, mass-energy, linear momentum and angular momentum* (spin). The application of these laws have led to the prediction of new particles *e.g.* neutrinos.

There are, however, certain modes of decay that are allowed by the above conservation laws, but they are never observed. For example, let us consider the possible modes of proton-decay :

$$p^+ \rightarrow e^+ + \pi^o$$

$$p^+ \rightarrow \mu^+ + \pi^o$$

These modes of decay, however, never occur in nature (proton is a stable particle). Therefore, it is natural to suppose that these are prevented by some unknown laws of conservation. Thus new laws of conservation have been deduced from a study of all possible types of particle decay. These are conservation laws of baryon-number *(B)*, electron lepton-number (*L*) and muon lepton-number *(M)*. These laws plus the old conservation laws help us to determine whether any given decay is capable of taking place or not.

To establish the new conservation laws, three new quantum numbers *B*, *L* and *M* have been introduced.

Conservation of Baryon Number

Each particle is assigned a baryon number *B*, such that

$B =+ 1$ for all baryons $(p, n, \Lambda^{\circ}, \Sigma^{+}, \Sigma^{\circ}, \Sigma^{-}, \Xi^{-}, \Xi^{\circ}, \Omega^{-})$

$B =-1$ for all antibaryons $\left(\overline{p}, \overline{n}, \overline{\Lambda}^{\circ}, \overline{\Sigma}^{+}, \overline{\Sigma}^{\circ}, \overline{\Sigma}^{-}, \overline{\Xi}^{-}, \overline{\Xi}^{\circ}, \overline{\Omega}^{-}\right)$

$B = 0$ for all others (photon, leptons, mesons). According to the law of conservation, the total value of *B* remains constant in any process involving elementary particles.

Conservation of Electron Lepton-number : There is an electron lepton number *L* for all particles, such that

$L = + 1$ for electron (e^-) and electron-neutrino (v_e),

$L = - 1$ for antielectron (e^+) and electron-antineutrino $(\overline{v}_e)$

$L = 0$ for all others.

Again, the total value of *L* remains constant in any decay or reaction of elementary particles.

Conservation of Muon Lepton-number : There is a muon lepton-number *M* for all particles, such that

$M = +1$ for muon (μ^-) and muon-neutrino (v_μ)

$M = -1$ for antimuon (μ^+) and muonantineutrino $\left(\overline{v}_\mu\right)$

$M = 0$ for all others.

The total value of M remains conserved in any process of elementary particles.

As example of the above laws, let us consider the decay of a free neutron:

$$n \rightarrow p + e^- + \bar{v}_e$$

For this B = 1 → 1 + 0 + 0

and L = 0 → 0 + 1 + (–1).

The total value of B before and after the decay is 1, and that of L is 0. Thus both B and L are conserved and hence the neutron decay does take place in nature. We know that a free neutron is unstable.

Stability of Proton : The laws of baryon and lepton number conservations are responsible of the stability of the proton. Let us consider a possible mode of a free proton decay :

$$p \rightarrow e^+ + \pi^\circ$$

For this $B = 1 \rightarrow$ 0 + 0

and $L = 0 \rightarrow$ (–1) + 0

The total values of both B and L are *not* the same before and after the decay, and so the decay is forbidden by both B and L conservations. Proton decay is really forbidden because there are no baryons of mass smaller than that of proton.

It is for the conservation of the lepton number that in beta-decays an electron is produced together with an antineutrino, and a positron is produced together with a neutrino.

Maximum available Energy for $\mu^- \rightarrow e^- + 2v$: The μ^- meson decays from rest. Therefore, the conservation of linear momentum requires that two particles must be ejected. The only particle observed is the electron. Since the charge is conserved, the second particle must be a neutrino. But since the spin of μ-meson is ½, and the sum of the spins of the electron and neutrino can only be 0 or 1 (but never ½), the conservation of spin (angular momentum) requires another particle of opposite spin ½ to be ejected at the same time. This must be an antineutrino, so that the equation would be

$$\mu^- \rightarrow \quad e^- + v + \overline{v}$$

$$\text{spins} \qquad \frac{1}{2} \rightarrow \quad \frac{1}{2} + 0$$

This raises another difficulty. If the neutrino and antineutrino are ejected simultaneously, they would annihilate in flight to produce two γ-photons ($v + \overline{v}\ 2\gamma$). This, however, is not found.

This shows that v and $\overline{v}$ do *not* form particle-antiparticle pair. Hence v and $\overline{v}$ must be the neutrinos of different species. Two types of neutrinos are therefore postulated : *e*-neutrinos and μ-neutrinos, and the decay equation should now be written as

$$\mu^- \rightarrow e^- + v_\mu + \overline{v}_e .$$

Now, taking $M_\mu = 207\ m_e$, the total available energy Q of the reaction is given by

$$207\ m_\theta \rightarrow \ m_e + 0 + 0 + Q,$$

because the mass of neutrinos is zero.

$$\therefore \qquad Q = 206\ m_e.$$

But m_e is equivalent to 0.51 MeV, therefore

$$Q = 206 \times 0.51 = 105 \text{ MeV}.$$

If this energy is equally divided among the three particles, the average energy of the electron is 35 MeV.

PROBLEMS

1. *An electron and its antiparticle are annihilated after collision. How much energy will be obtained ? (Mass of electron =9.1 × 10^{-31} kg, speed of light = 3.0 × 10^8 m/s and Planck's constant = 6.6 × 10^{-34} J-s).*

Solution: The positron is antiparticle of electron. In electron-positron collision, both are annihilated and the total mass is converted into two γ-photons (energy) :

$$e^- + e^+ \rightarrow 2\gamma.$$

The mass-loss in this process is

$$\Delta m = 2 \times (9.1 \times 10^{-31}) = 18.2 \times 10^{-31} \text{ kg}.$$

According to mass-energy relation, the energy released is

$$\Delta E = (\Delta m)c^2$$

$$= (18.2 \times 10^{-31} \text{ kg}) \times (3.0 \times 10^8 \text{ m/s})^2$$

$$= 1.64 \times 10^{-13} \text{ joule}$$

$$= \frac{1.64 \times 10^{-13}}{1.6 \times 10^{-19}} \; eV$$

[because 1 eV = 1.6 × 10^{-19} joule]

$$= 1.02 \times 10^6 \text{ eV} = 1.02 \text{ MeV}.$$

This is total energy obtained.

Two γ-photons are produced in this process, hence energy of each photon

$$E = \frac{\Delta E}{2} = \frac{1.64 \times 10^{-13} \text{ joule}}{2}$$

$$= 0.82 \times 10^{-13} \text{ joule} = 0.51 \text{ MeV}.$$

2. *A positron and an electron with negligible kinetic energy annihilate each other to produce two photons. What are their frequencies.*

Solution: According to the above example, the energy of each of the two photons produced by the union of an electron and a positron is

$$E = 0.82 \times 10^{-13} \text{ joule.}$$

If the frequency of the photon be v, then $E = hv$.

$$\therefore \qquad v = \frac{E}{h} = \frac{0.82 \times 10^{-13} \text{ joule}}{6.6 \times 10^{-34} \text{ joule} - \text{second}}$$

$$= 1.2 \times 10^{20} \text{ hertz.}$$

3. ***If a pion decays from rest to give a muon of 4.0 MeV energy, what is the kinetic energy of the accompanying neutrino? What is the mass of the neutrino in this process ?***

Solution: The pion decay is

$$\pi^+ \rightarrow \mu^+ + v_\mu + Q\ .$$

From this, we have

$$Q = M_\pi - M_\mu \quad \text{(neutrino has zero 'rest' mass)}$$

$$= 273\ m_e - 207\ m_e$$

$$= 66\ m_e$$

$$= 66 \times 0.51 \qquad (\because \ m_e \equiv 0.51 \text{ MeV})$$

$$= 33.7 \text{ MeV.}$$

Since the muon takes 4.0 MeV of energy, the energy shared by the neutrino is 33.7 – 4.0 = 29.7 MeV.

Mass of this neutrino

$$= \frac{29.7}{0.51}\ m_e = 58.2\ m_e.$$

4. ***Find the maximum kinetic energy of the electron emitted in the beta decay of the free neutron. The neutron-proton mass difference is 1.30 MeV.***

Solution: The decay equation is

$$n \rightarrow p^{+} + e^{-} + \bar{v}_e + Q.$$

From this, we have

$$\begin{aligned} Q &= M_n - M_p - m_e \\ &= (1.30 \text{ MeV}) - m_e \\ &= (1.30 \text{ MeV}) - (0.51 \text{ MeV}) \\ &= 0.79 \text{ MeV}. \end{aligned}$$

This energy-release goes into the kinetic energy of the electron plus the kinetic energy of the antineutrino. The two particles can share the 0.79 MeV of energy in any way they choose. Thus the electron may have a maximum kinetic energy of 0.79 MeV.

3

The Accelerators

Purpose of Accelerators

Particle accelerators are machines for imparting high energies of the order of MeV, to more, to charged particles (electrons, protons, deuterons, α-particles), and also to bare nuclei of lighter atoms. Such energetic particles are required in making nuclear investigations, and in the production of transuranic elements.

There are three main types of particle accelerators : (*i*) electrostatic accelerators (such as Van de Graff generator); (*ii*) linear accelerators; and (*iii*) orbital accelerators (such as cyclotron, synchro-cyclotron, betatron, and the synchrotron).

Linear Accelerator

A linear accelerator can impart energies to heavy ions, protons and electrons up to more than 10 MeV, which is the upper limit attainable by an electrostatic accelerator. In this accelerator the particle receives a series of additive accelerations in successive stages.

Construction : The principle diagram of an accelerator is shown in Fig. It consists of a long, evacuated tube carrying a series of coaxial cylindrical electrodes, 1, 2, 3, 4,..., of successively increasing length. Alternate sets of cylinders are connected together, and a high-frequency alternating p.d. is applied between the two sets by a radio-frequency oscillator. Thus alternate cylinders have opposite potentials. For example, in one half-cycle, cylinders, 1, 3, 5, ... are positive whilst 2, 4,,.. are negative; whereas in the next half-cycle 1, 3, 5,... become negative whilst 2,4, ...become positive.

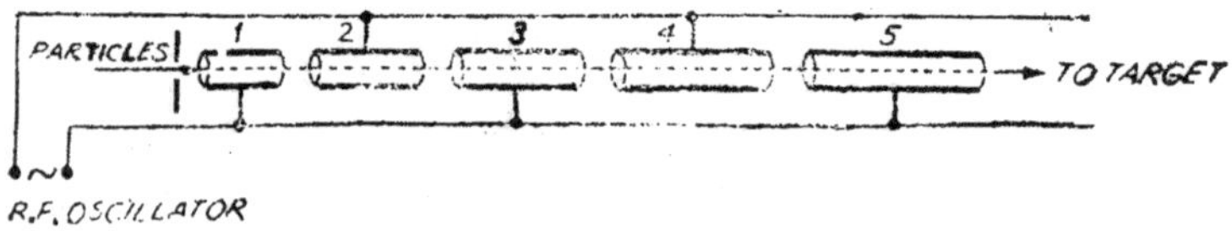

Working : A charged particle keeps a constant speed while moving in the field-free space *within* a cylinder, but is accelerated while moving across the gap between two cylinders having a p.d.

Suppose that a positive particle enters cylinder 1 at a moment when the cylinders 1, 3, 5,... just acquire a negative potential, and the cylinders 2, 4,... a positive potential. The particle travels inside the cylinder with a constant speed. The frequency of the alternating p.d. applied between the two sets of cylinders is so chosen that the time taken by the particle to move through the cylinder is equal to the time of one half-cycle of the p.d. Thus as soon as the positive particle leaves the cylinder 1, the cylinders 1, 3, 5,... become positive and the cylinders 2, 4,... negative. The particle is therefore accelerated in the gap between cylinders 1 and 2, and enter the cylinder 2 with an increased speed. Once inside the cylinder 2, it travels with constant increased speed remaining unaffected by the changing potential of the cylinder. Since the successive cylinders are made longer, the particle takes the same time to traverse each cylinder in spite of its increased speed. Therefore, by the time the (positive) particle leaves the cylinder 2, the cylinders 2, 4,.... become again positive and the cylinders 1, 3, 5,... again negative. The particle is therefore again accelerated in the gap between 2 and 3 and reaches the cylinder 3 with a further increase in speed. The same happens at each successive gap right down the tube, and the particle goes on gaining energy. There is no theoretical limit to the final energy which is limited only by the length of the surrounding tube and the available power.

A typical length for a proton accelerator is between 5 and 30 meters. It can be fed with 4-5 MeV protons already accelerated by a Van de Graff generator, and can further increase their energy to about 50 MeV.

A linear accelerator can also impart energies to electrons. An electron, however, has a speed as high as 98% of the speed of light at an energy of only 2 MeV. The transit times through cylinders are therefore so short that extremely high frequency p.d.'s ($\approx$ 1000 Mc/s) are required. Furthermore, beyond 2 MeV any further increase in electron speed is hardly possible, and any gain in energy comes rather from the relativistic *increase in mass* of the electron. The world's longest linear accelerator, about 3 km in length, delivers electrons with an energy of 30,000 MeV or more.

Cyclotron

A linear accelerator is inconvenient in shape. A more compact machine called a 'cyclotron' was devised by Lawrence and Livingston in 1932 to accelerate positive ions such as protons, deuterons and α-particles.

Constriction : It consists of two hollow D-shaped copper electrodes D_1 and D_2 (called the 'dees') with a small gap between them (Fig.). The dees are enclosed in an evacuated steel box placed between the poles of a very large magnet. The magnet produces a field B of about 15,000 gauss perpendicular to the plane of dees. A high-frequency alternating p.d. ($\approx$ 12 Mc/sec $\approx 10^5$ volts) is applied across the dees by means of a powerful valve oscillator.

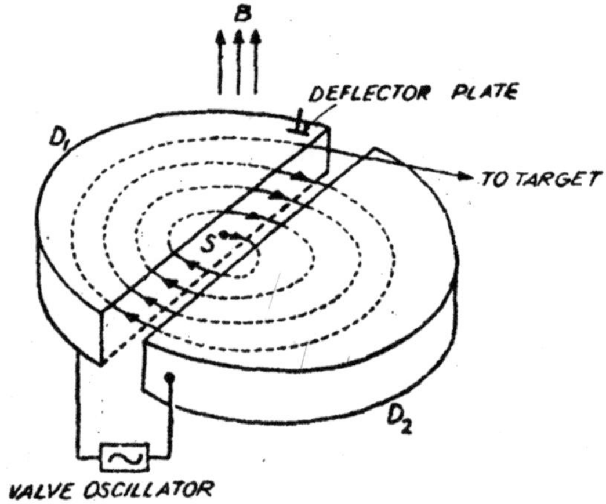

The ion source consists of a small chamber carrying a heated filament and the gas concerned (hydrogen for protons, heavy hydrogen for deuterons). The electrons emitted by the filament ionise the gas. The ions so produced pass through a capillary tube which ejects them at the centre 5 of the gap between the dees.

Working : Suppose that a positive ion is injected at S at a moment when D_2 just acquires a negative potential and D_1 a positive potential. The ion is then accelerated towards D_2 and enters it with increased velocity. Once inside D_2 it becomes free from the electric field and travels with constant speed. Since the ion is moving in a magnetic field B perpendicular to the plane of dees, it adopts a circular path of radius r given by

$$\frac{mv^2}{r} = Bqv.$$

where m is the mass, q the charge and v the speed of the ion.

$$\therefore \qquad v = \frac{Bqr}{m}. \qquad \ldots (i)$$

The time T required by the ion to complete one circular revolution is $2\pi r/v$. Therefore

$$T = \frac{2\pi r}{v} = \frac{2\pi m}{Bq}$$

The frequency of revolution is

$$f = \frac{1}{T} = \frac{Bq}{2\pi m}. \qquad (ii)$$

Thus, if B, m and q are constants, the period of revolution T, or the frequency, is independent of the radius of the path and of the speed of the ion. That is, all ions (same m, same q) 'take the same time to describe a circle (or semi-circle) irrespective of their speeds or energies. (In fact, it is this result which makes the cyclotron possible).

The frequency of the applied p.d. is adjusted to be equal to the frequency of revolution of the ions. When this is the case, the positive ion moving in D_2 completes a semi-circle and leaves it just at the moment when D_2 becomes positive and D_1 negative. The ion is therefore further accelerated in the gap and enters D_1. On account of its increased speed its semi-circular path in D_1 is of greater radius. *The time of passage through D_1, however, is still the same.* The process is repeated after every half-cycle of the p.d., and the ion gains speed each time it passes from one dee to the other. Finally, the ion reaches the outer edge of one dee where it is pulled out of the system by a negatively-charged deflector plate.

Evidently, eq. (*ii*), when f is taken as the frequency of the applied p d., is the resonance condition for the successful operation of the cyclotron.

The cyclotron has produced positive particles having energies of 20 MeV.

Attained Energy

The velocity, and hence the kinetic energy, with which the ion leaves the cyclotron can be obtained by putting $r = R$ (radius *of* dees) in eq (*i*). Thus

$$v = \frac{BqR}{m}$$

Therefore, the kinetic energy of the emerging ion is

$$K = \frac{1}{2}mv^2 = \frac{1}{2}m\left(\frac{BqR}{m}\right)^2$$

$$= \frac{1}{2}m\ B^2R^2\left(\frac{B}{m}\right)^2.$$

But $f = \dfrac{Bq}{2\pi m}$, where f is the frequency of the applied p.d.

Therefore $K = 2\pi^2 f^2 m R^2.$

This is the expression for the energy of the particle in terms of the frequency of applied p.d. and the radius of the dees. The frequency, in turn, depends upon the flux density B. Thus in the non-relativist range the limitation on the maximum energy attained by a particular ion is determined by the size of the dees and the pole-faces of the magnet.

Limitation : The maximum energies of particles attainable in a cyclotron cannot be increased indefinitely by increasing the size of the apparatus because of the relativistic increase in mass of the particles in accordance with the equation

$$m = \frac{m_0}{\sqrt{\left(1-\frac{v^2}{c^2}\right)}},$$

where m_0 is the rest mass of the particle and in the mass at speed v. As the speed v of the particle increases to become comparable with the speed of light c, its mass m increases appreciably. Therefore the frequency of revolution f $\left(\frac{Bq}{2\pi m}\right)$ goes on decreasing. Thus the particle traverses each dee too slowly and becomes more and more out of step with the applied p.d. until it can no longer be accelerated further.

At 20 MeV the mass of a proton is about 2% greater than its rest mass, but beyond this the increase in mass becomes serious enough to affect the operation of the cyclotron. Thus it is the upper limit of proton energy attainable by a cyclotron.

The electrons due to their very small mass approach the speed of light very quickly so that the relativistic increase in mass becomes appreciable at low enough energies. For example, the mass of an electron is doubled at an energy of only about 0.5 MeV. Hence the cyclotron *cannot* be used to accelerate electrons even to moderate energies.

Overcoming of Limitation (Synchro-cyclotron) : The energy limitation of 20 MeV resulting from the relativistic increase m the mass of the particle has been removed in a modified form of the cyclotron known as 'synchro-cyclotron' (or 'frequency-modulated cyclotron'). In this machine a high-frequency p.d. is applied at the moment of injection ; but as the frequency of revolution of the particle begins to slow down, the frequency of the p.d. is also reduced so that synchronism between the revolving particle and the accelerating p.d. is maintained. Using this technique, protons of 300 MeV energy and α-particles of 384 MeV energy have been obtained. The maximum energy is decided by the size of the magnet.

Betatron : The betatron is used to accelerate electrons to energies higher than the electrons (β-particles) emitted from radioactive substances. The betatron-accelerated electrons can produce X-rays so short in wavelength and so penetrating in character as to surpass the γ-rays from radioactive substances.

Principle : In the betatron the electrons are put in a *stable* circular orbit in which they are continuously accelerated due to an emf produced by an *increasing* magnetic flux through the orbit.

Let us consider an electron moving in a circular orbit of radius r in a magnetic field at right angles to the plane of the orbit (Fig.). The magnetic field is *non-uniform* in space, and *increasing* with time. Suppose, at a given instant, $\overline{B}$ is the average flux density *through* the orbit and B, the flux density *at the* orbit itself. The instantaneous magnetic flux Φ through the orbit will be given by

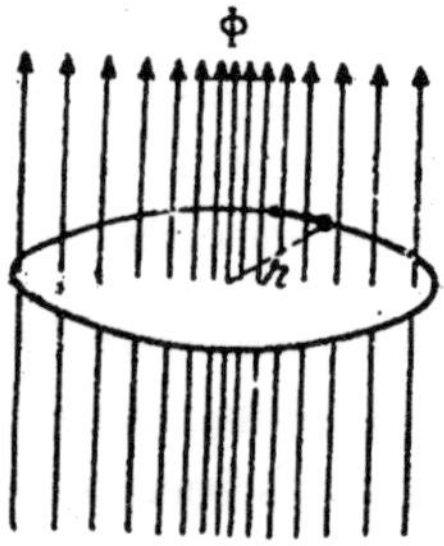

Φ = average flux density through the orbit × area enclosed by the orbit

$$= \overline{B} \times \pi r^2 = \pi r^2 \overline{B} \qquad \text{... (i)}$$

As the magnetic field is increasing with time, the magnetic flux Φ through the orbit is increasing at a rate $\frac{d\Phi}{dt}$. As a result, an emf is induced in the orbit. If V is the induced emf, then

$$V = \frac{d\Phi}{dt} \text{ (numerically).}$$

The work done on the electron in one complete revolution is

$$W = \text{charge on electron} \times \text{emf}$$

$$= eV.$$

Hence the tangential force acting on the electron is

$$F = \frac{\text{work done in one rev.}}{\text{distance moved in one rev.}}.$$

$$= \frac{W}{2\pi r} = \frac{eV}{2\pi r} = \frac{e}{2\pi r}\frac{d\Phi}{dt}.$$

Substituting the value of Φ from eq. (*i*), we obtain

$$F = \frac{e}{2\pi r}\frac{d\left(\pi r^2 \overline{B}\right)}{dt} = \frac{er}{2}\frac{d\overline{B}}{dt}. \qquad \text{... } (ii)$$

The electron is accelerated by this force acting upon it. This force arises due to the emf produced by the *increasing* 'average flux density $\overline{B}$ through the entire orbital area. As $\overline{B}$ increases the electron energy increases and the electron tends to move it in an orbit of *increasing* radius.

Now, the electron is moving in a circular orbit of radius r in a magnetic field perpendicular to it. The instantaneous magnetic flux density at the orbit is B_r. If v be the instantaneous speed of the electron, the instantaneous radial force on it will be $B_r ev$. This must be equal to the centripetal force mv^2/r. Thus

$$Brev = \frac{mv^2}{r}.$$

Therefore, the instantaneous momentum of the electron is

$$p = mv = B_r er.$$

The rate of change of momentum is

$$\frac{dp}{dt} = \frac{d(brer)}{dt} = er\,\frac{dBr}{dt}.$$

By Newton's law this must be equal to a tangential force F acting on the electron. Thus

$$F = er\frac{dBr}{dt}. \qquad \text{... } (iii)$$

Clearly this force arises by the *increasing* flux density B_r at the orbit. As B_r increases, the electron tends to move in an orbit of *decreasing* radius ($B_r\, ev = mv^2/r$).

Thus the forces given by eq. (*ii*) and (*iii*) have opposite actions on the radius of the orbit ; the former tends to increase while the latter tends to decrease the radius Hence, to preserve a constant radius of the orbit these two forces must be equal numerically. That is

$$\frac{er}{2}\frac{d\overline{B}}{dt} = er\,\frac{dB_r}{dt}.$$

Integrating both sides of the equation, assuming that $\overline{B} = B_r = 0$ at $t = 0$, we obtain

$$\frac{er\overline{B}}{2} = erB_r$$

or $$\overline{B} = 2B_r. \qquad \text{(Betatron condition)}$$

Thus the average magnetic flux density through the orbit must always be twice the flux density at the orbit. This is the condition of betatron. This condition is achieved by suitably designing the size and shape of the magnet providing the magnetic field. When this is done, the electrons once put in the orbit go again and again around the same orbit with continuous increasing speed so long as the magnetic field is increasing (with time).

In betatron the increase in speed around the circular orbit is *continuous* and there is no synchronism to be maintained. Hence any relativistic increase in mass of the electron and the consequent slowing down of the rate of revolution do not affect the operation of the betatron.

Construction : The construction of the betatron is illustrated in Fig. It consists of a" large highly-evacuated annular tube *DD*, called a "donut" tube, in which the electrons revolve.

Their orbit is a circle at right angles to the plane of the page. They are shown entering the page at the right (e) and emerging the page at the left (x). The tube *DD* is placed between the specially-designed pole-pieces of a large electromagnet energised by an alternating current. The magnet satisfies the betatron condition by providing a larger flux density in the central part of the field. Since the magnet is energised by *alternating* current, an increasing magnetic field in a given direction is obtained only during one-quarter cycle in which the current increases from zero to the peak value. Hence the electrons are kept in their orbit only during this part of each cycle, and the output electron beam is in pulses.

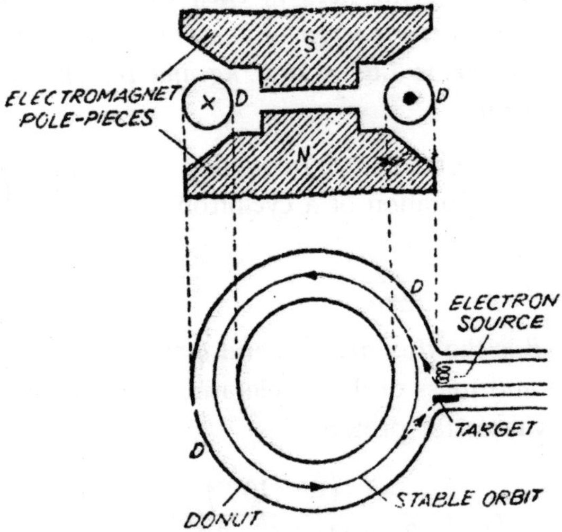

Action : The plane of the orbit is shown in the diagram. The electrons to be accelerated are injected into the donut tube *D* by means of an electron-gun. The gun is so *pulsed* that the electrons are injected only at the start of a quarter-cycle in which the magnetic field increases in a given direction from zero to the peak value. These electrons are soon captured into a *stable* orbit and make several hundred thousand revolutions during the quarter-cycle, being accelerated continuously. At the end of each accelerating quarter-cycle the orbit is momentarily expanded by means of an auxiliary magnetic field (produced by sending a pulse of an extra current through an auxiliary coil), so that the electrons strike the target or emerge through a window.

Betatrons have produced electrons with energies of greater than 300 MeV.

PROBLEMS

1. A cyclotron in which the magnetic flax density is 3 weber/ m^2 and whose dees' radius is 0.5 m, is used to accelerate (i) protons, (ii) α-particles. What must be the frequency of the p.d. applied across the dees in each case ? Also calculate the maximum kinetic energy of the emerging particles. The mass of a proton is 1.67×10^{-27} kg and the charge on it is 1.6×10^{-19} coul.

Solution: The resonance frequency of the applied p.d. for the successful operation of a cyclotron is

$$f = \frac{Bq}{2\pi m}, \qquad \text{... } (i)$$

where B is the flux density, and q and m are respectively the charge and mass of the accelerating particle. Substituting the given values for proton

$$f_p = \frac{3 \times \left(1.6 \times 10^{-19}\right)}{2 \times 3.14 \times 1.67 \times 10^{-17}}$$

$$= 46 \times 10^6 \text{ cycles/sec}$$

$$= 46 \text{ mc/sec.}$$

The mass of an α-particle is 4 times that of proton, while the charge is twice that of proton. Hence, by eq. (i), the required frequency is

$$f\alpha = \frac{1}{2} f_p = 23 \text{ mc/sec.}$$

To find the maximum energy of the emerging particles let us consider the relation

$$\frac{mv^2}{r} = Bqv$$

so that $$v = \frac{Bqr}{m}$$

If $r = R$, the radius of dee, then v would represent the (maximum) velocity of the emerging particle. Thus, the maximum kinetic energy is

$$K = \frac{1}{2} mv^2 = \frac{1}{2} m \left(\frac{Bqr}{m}\right)^2$$

$$= \frac{1}{2} \frac{B^2 q^2 R^2}{m}. \quad \text{... } (ii)$$

Substituting the given values for protons,

$$K_p = \frac{1}{2} \times \frac{(3)^2 \times \left(1.6 \times 10^{-19}\right)^2 \times (0.5)^2}{1.67 \times 10^{-27}}$$

$$= 1.72 \times 10^{-11} \text{ joule.}$$

Now $1 \text{ eV} = 1.6 \times 10^{-19}$ joule.

$$\therefore \quad K_p = \frac{1.72 \times 10^{-11}}{1.6 \times 10^{-19}} = 1.1 \times 10^8 \text{ eV} = 110 \text{ MeV}.$$

It can be seen from eq. (*ii*) that the kinetic energy of the emerging α-particles (charge double, mass four times than proton) would be the same as for protons. Thus

$$K_\alpha = K_p = 110 \text{ MeV}.$$

2. A cyclotron is being used for accelerating α-particles. The frequency of potential on the dees for α-particles has been 16 mc/sec and maximum energies of 48 MeV have been attained. What change in frequency, if any, is needed to accelerate

deuterons and what maximum energies may be attained ? (The mass, as well as the charge, of a deuteron is half of that of an α-particle).

[**Ans.** No change in frequency, 24 MeV]

3. The diameter of a cyclotron is 18 m and the strength of magnetic field is 0.7 weber/m². Calculate the energies to which (i) a proton, (ii) a deuteron can be accelerated. Mass of a proton, m_p= 1.008 amu ; charge of an electron, e=1.6 × 10^{-19} coulomb and 1 amu = 1.66 × 10^{-27}kg.

Solution: When the particle describes a circle of radius *R*, equal to Dees' radius, the velocity reaches maximum value. Equating the magnetic force to the centripetal force, we have

$$Bqv = \frac{mv^2}{R}$$

or

$$v = \frac{BqR}{m}.$$

∴ the (maximum) kinetic energy is

$$K = \frac{1}{2}mv^2$$

$$= \frac{B^2q^2R^2}{2m} \quad \text{... (i)}$$

The charge on a proton is + 1.6 × 10^{-19} coul (equal to the negative charge of electron) and its mass is 1.008 × (1.66 × 10^{-27} kg) = 1.67 × 10^{-27} kg. Therefore, the maximum kinetic energy of the proton is

$$K_p = \frac{\left(0.7 \text{ weber}/\text{m}^2\right)^2 \left(1.6\times 10^{-19} \text{ coul}\right)^2 \left(0.9 \text{ m}\right)^2}{2\times 1.67\times 10^{-27} \text{ kg}}$$

$$= 3.0 \times 10^{-12} \text{ joule}$$

$$= \frac{3.0\times 10^{-12}}{1.6\times 10^{-19}} = 1.9 \times 10^7 \text{ eV} = 19 \text{ MeV}.$$

The charge on deuteron is same as on proton, but its mass is almost double than that of proton. Hence, as seen from eq. (*i*) its kinetic energy will be almost half than that of proton.

$$\therefore \qquad K_\alpha = 9.5 \text{ MeV}.$$

4. *A cyclotron with a radius of 10 m has a magnetic field of 2.0 T (tesla). Determine the maximum energy of the accelerated deuterons. Through what potential difference would they have to travel to attain the energy ? The mass of deuteron is 3.34 × 10^{-27} kg and electronic charge is 1.6 × 10^{-19} coulomb.*

Solution: The maximum kinetic energy of the deuterons is given by

$$K = \frac{B^2 q^2 R^2}{2m}$$

Substituting the given values, remembering that the positive charge on deuteron is equal to the negative charge on electron, we get

$$K = \frac{(2.0 \text{ tesla})^2 (1.6 \times 10^{-19} \text{ coulomb})^2 (1.0 \text{ m})^2}{2 \times 3.34 \times 10^{-27} \text{ kg}}$$

$$= 1.53 \times 10^{-11} \text{ joule}$$

$$= \frac{1.53 \times 10^{-11}}{1.6 \times 10^{-19}} = 0.96 \times 10^8 \text{ eV} = 96 \text{ MeV}.$$

To achieve this energy, the deuteron will have to travel through a p.d. of 96 million volts.

5. *Calculate the maximum energy which a cyclotron can give to α-particles if the diameter of dee is 1.2 meter and the magnetic field is 1.5 wb/m². At what frequency the cyclotron be operated to achieve this energy ? The mass of an α-particle is 6.645 × 10^{-27} kg and the charge on it is 3.2 × 10^{-19} coulomb. (1 MeV = 1.6 × 10^{-13} joule).*

[**Ans.** 39 MeV, 11.5 Mc/sec.]

6. A cyclotron whose magnetic pole faces have a diameter of 1.52 meters is operated with a radio-frequency field at 10^7 cycles per sec. Show that it can accelerate protons to an energy of approximately 12 MeV and α-particles to approximately 48 MeV. The mass of a proton is 1.67×10^{-27} kg and that of an α-particle roughly four times. ***[Hint : $K = 2\pi^2 m f^2 R^2$***

7. A cyclotron of diameter 1 meter is used to accelerate protons. Across it is applied an alternating voltage of 10 mc/sec frequency and of peak value 5000 volts. Calculate the magnetic field needed and the energy of the emergent beam. The mass and charge of proton are 1.67×10^{-27} kg and 1.6×10^{-19} coulomb respectively.

Solution: The resonance frequency of the applied voltage for the successful operation of the cyclotron is

$$f = \frac{Bq}{2\pi m'}$$

so that the required field is

$$B = \frac{2\pi mf}{q}$$

Substituting the given values :

$$B = \frac{2 \times 3.14 \times \left(1.67 \times 10^{-27} \text{ kg}\right) \times \left(10 \times 10^{6} \text{ sec}^{-1}\right)}{1.6 \times 10^{-19} \text{ coul}}$$

$$= 0.66 \text{ weber/meter}^2.$$

The kinetic energy of the emerging proton is

$$K = \frac{B^2 q^2 R^2}{2m}$$

$$= \frac{\left(0.66 \text{ weber/m}^2\right)^2 \times \left(1.6 \times 10^{-19} \text{ coul}\right)^2 \times \left(0.5 \text{ meter}\right)^2}{2 \times \left(1.67 \times 10^{-27} \text{ kg}\right)}$$

$$= 8.0 \times 10^{-13} \text{ joule} = \frac{8.0 \times 10^{-13}}{1.6 \times 10^{-13}} = 5 \text{ MeV}.$$

8. Calculate the energy of a proton when it emerges out of a cyclotron after 50 revolutions. The voltage applied across the dees is 20 kV.

Solution: The proton (charge *e*) passes across the gap between the two dees twice in each revolution, and acquires energy *eV* during each transit, where *V* is the peak value of the voltage applied. Thus the energy acquired in 50 revolutions is

$$K = 2 \text{ eV} \times 50 = 100 \, eV$$

$$= 100 \times (1.6 \times 10^{-19} \text{coul}) \times 20{,}000 \text{ volts}$$

$$= 3.2 \times 10^{-13} \text{ joule.}$$

$$\text{Now, } 1 \text{ MeV} = 1.6 \times 10^{-13} \text{ joule.}$$

$$\therefore \quad K = \frac{3.2 \times 10^{-13}}{1.6 \times 10^{-13}} = 2 \text{ MeV.}$$

9. ***If the maximum r. f. voltage applied to a cyclotron be 10 kV, show that a proton has to make 900 revolutions within the cyclotron before achieving $\frac{1}{5}$th the velocity of light. Take proton mass 1.6×10^{-27} kg and charge 1.6×10^{-19} coulomb. The velocity of light is 3×10^{8} m/sec.***

10. ***An alternating potential difference of 50 kV maximum value is applied to the dees, whose radius is 40 cm, of a cyclotron by an oscillator. A deuteron of mass 2 amu acquires energy of 4 MeV in the cyclotron. Calculate (i) magnetic field strength, (ii) frequency of oscillator, (iii) number of revolutions which the deuteron has to make inside the cyclotron to gain energy. (1 amu= 1.6×10^{-27} kg, $e = 1.6 \times 10^{-19}$ coulomb).***

Solution: The expression for the kinetic energy of particles of mass *m* and charge *q* emerging from a cyclotron is

$$K = \frac{B^2 q^2 R^2}{2 \text{m}},$$

where *B* is the magnetic field and *R* the dees radius.

Here $q = e = 1.6 \times 10^{-19}$ coul,

$R = 0.4$ meter,

$m = 2\text{amu} = 2 \times 16 \times 10^{-27}$ kg,

and $K = 4\text{MeV} = 4 \times 1.6 \times 10^{-13}$joule.

Thus

$$B^2 = \frac{2\,Km}{q^2R^2}$$

$$= \frac{2 \times \left(4 \times 1.6 \times 10^{-13}\right) \times \left(2 \times 1.6 \times 10^{-27}\right)}{\left(1.6 \times 10^{-19}\right)^2 \times (0.4)^2} = 1.0.$$

$\therefore$ $B = 1.0$ weber/m^2.

The (resonance) frequency of p. d. required for the successful operation of the cyclotron is

$$f = \frac{Bq}{2\,\pi\, m} = \frac{1.0 \times \left(1.6 \times 10^{-19}\right)}{2 \times 3.14 \times \left(2 \times 1.6 \times 10^{-27}\right)}$$

$$= 8 \times 10^6 \text{ cycles/sec} = 8 \text{ mc/sec.}$$

Suppose the deuteron makes n revolutions to gain energy K. In each revolution it is pushed twice and in each push gains energy qV, where V is the peak value of the applied alternating voltage. Thus

$$n \times 2qV = K$$

$$\therefore \quad n = \frac{K}{2\,qV}$$

$$= \frac{4 \times 1.6 \times 10^{-13} \text{ joule}}{2 \times \left(1.6 \times 10^{-19} \text{ coul}\right) \times (50{,}000 \text{ volts})} = 40.$$

11. In a certain betatron the maximum magnetic field at the orbit was 0.4 tesla, operating at 50 cps with a stable orbit

diameter of 1.5 meter. Calculate the average energy gained per revolution and the final energy of the electrons. Assume that the electrons move with nearly the velocity of light.

Solution: In betatron the electrons move with velocity nearly c so that the total distance travelled in the acceleration time (one-quarter cycle) is

$$d = c \times \frac{T}{4} = \frac{c}{4f},$$

where f is the frequency of the A.C. supply feeding the magnet. If r be the radius of the stable orbit, then the total number of revolutions made by the electrons in the acceleration time is

$$n = \frac{d}{2\pi r} = \frac{c}{8\pi f r}.$$

Substituting the given values :

$$n = \frac{3 \times 10^8 \text{ m/sec}}{8 \times 3.14 \times 50/\text{sec} \times 0.75 \text{ m}} = 3.17 \times 10^5.$$

Relativistically, the momentum of the electron is given by

$$p = \frac{E}{c}$$

where E is the final energy acquired. If B is the maximum magnetic field *at the orbit*, and v is the final velocity acquired, then we have

$$Bev = mv^2/r,$$

where m is the (relativist) mass of the electron at velocity v. Thus

$$mv = Ber$$

or $$p = mv = Ber = \frac{E}{c}$$

or $$E = Berc.$$

Substituting the given values, the final energy acquire is

$$E = (0.4 \text{ weber/m}^2) \times (1.6 \times 10^{-19} \text{ coul}) \times (0.75\ m) \times (3 \times 10^8 \text{ m/sec})$$

$$= 1.44 \times 10^{-11} \text{ joule}$$

$$= \frac{1.44 \times 10^{-10}}{1.6 \times 10^{-13}}$$

$$= 90 \text{ MeV.} \quad [\therefore 1 \text{ Mev} = 1.6 \times 10^{-13} \text{ joule}]$$

The average energy per revolution is

$$\frac{90 \times 10^6 \text{ eV}}{3.17 \times 10^5} = 284 \text{ eV.}$$

12. ***Calculate average energy per revolution and final energy gained by electrons in a betatron to which is applied a maximum magnetic field of 5000 gauss operating at 60 hertz in a stable orbit of diameter 2 meters. (1 gauss=10^{-4} tesla, $c = 3 \times 10^8$ m/sec, $e = 1.6 \times 10^{-19}$ coul and 1 MeV = 1.6×10^{-13} joule).***

[Ans. 750 eV, 750 MeV]

4

Decay of Radioactive Nucleus

Explanation of α-emission from Radioactive Nuclei : The nuclei of heavier atoms, beyond bismuth ($_{83}Bi^{209}$), are unstable (radioactive) with respect either to α- or to β-emission. This is because these nuclei are so large that the *short-range* nuclear forces holding the nucleons together are hardly able to counterbalance the electrostatic repulsion between the large number of protons in them, α-emission occurs in such nuclei as a means of increasing their stability by reducing their size. In fact, all nuclei with $Z > 83$ and $A>209$ spontaneously decay into lighter nuclei through the emission of α-particle which is $_2He^4$ nucleus. The equation for α-decay can be written as :

$_ZX^A$	$\rightarrow$	$_{Z-2}Y^{A-4}$	+	$_2He^4$
Parent nucleus		Daughter nucleus		Alpha particle

Since each α-particle is a helium nucleus composed of two protons and two neutrons, their emission from nuclei fits with the proton-neutron picture of the nucleus.

An important question is that why do the radioactive nuclei emit α-particles ($_2He^4$) rather than protons ($_1H^1$) themselves. The answer lies in the *high* binding energy of the α-particle. To escape from a nucleus, a particle must have kinetic energy. Only the α-particle mass is sufficiently smaller than the sum of the masses of its constituent nucleons. Therefore, in the formation of α-particle within the nucleus sufficient energy is released which becomes available to the particle to escape.

Again, there remains the problem of *how* an α-particle can actually escape the nucleus. A nucleus is surrounded by a potential barrier, and the escaping particle must have enough energy to cross the barrier. The uranium nucleus, for α-particles, has a potential barrier of 27 MeV height so that only particles having 27 MeV or more energy would be able to escape. But α-particles emitted by uranium have an energy of only 4 MeV. Then how they get out across the barrier at all ? The explanation is quantum-mechanical and is based on the following lines :

(i) An α-particle, before emission, exists as such within the nucleus.

(ii) It is in constant motion but is kept inside the nucleus by the surrounding potential barrier.

(iii) There is a small, but finite, probability that the particle may "leak" through the barrier each time it collides with it. (This is due to the wave nature of the particle).

(iv) Once the particle leaks through the barrier, it escapes from the nucleus because of its kinetic energy and the electrostatic repulsion.

An α-particle within a nucleus collides with the surrounding barrier about 10^{21} times per second. Calculations show that in case of U^{238} nucleus the chance of escape in any single collision is only 1 out of 10^{38}. Thus an α-particle may have to try for $\frac{10^{38}}{10^{21}} = 10^{17}$ sec $= 3 \times 10^9$ years before it actually escapes. This explains the very long half-life of U^{238} which runs in billions of years (4.5 billions).

Po^{214}, in contrast to U^{238}, has a half-life of only 0.0001 sec. There are two reasons for it : (i) An α-particle in Po^{214} nucleus has a large energy. (*ii*) The height of the potential barrier is smaller. Therefore the chance of escape in a single collision is much larger, 1 out of 10^{17}. Hence the mean waiting time for the escape reduces to $\frac{10^{17}}{10^{21}} = 0.0001$ sec.

Explanation of β-emission from Radioactive Nuclei : Like α-decay, the β-decay and positron emission are the means by which a nucleus alters its composition (neutron/proton ratio) to achieve greater stability.

β-decay is the mission of electrons from nuclei. All properties of nuclei, however, firmly indicate that they do not contain electrons. They are composed of protons and neutrons only. Then how electrons are emitted from radioactive nuclei? Yukawa's meson theory supplies an answer to this question.

In the nucleus the protons and neutrons are held together by the continual exchange of pi-mesons between them. The electron emission takes place on the spontaneous conversion of a neutron into a proton by ejecting a π^--meson. This π^--meson decays almost instantly into an electron (e^-) and an antineutrino ($\bar{\nu}$). The reaction is

$$n \rightarrow p + \bar{\pi} \rightarrow p + e^- + \bar{\nu}.$$

Since in β-decay a neutron is converted into a proton, the neutron/ proton ratio decreases.

In β-decay, the mass number (number of nucleons) of the radioactive nucleus remains unchanged, but the atomic number (number of protons) increases by unity. Hence the equation for β-decay can be written as

$_ZX^A \rightarrow$	$_{Z+1}Y^A +$	$_{-1}\beta^o + \bar{v}$.
Parent nucleus	Daughter nucleus	Beta particle (electron)

Similarly, the positron emission takes place on the spontaneous conversion of a proton into a neutron by ejecting π^+-meson which decays almost instantly into a positron (e^+) and a neutrino. The reaction is

$$p \rightarrow n + \pi^+ \rightarrow n + e^+ + v.$$

In this process the neutron/proton ratio increases.

In positive β-decay (positron emission) the mass number remains unchanged, but the atomic number decreases by unity. Hence the equation for positron emission can be written as

$_ZX^A \rightarrow$	$_{Z-1}Y^A +$	$_{+1}\beta^o + v$.
Parent nucleus	Daughter nucleus	Positron

The significance of the emission of antineutrino in β-decay and of neutrino in positron emission is that it keeps energy, linear momentum and angular momentum all conserved.

Electron Capture

The electron capture is a process which is competitive with positron emission. In this process, a nucleus captures one of the inner orbital electrons of the atom, with the result that a nuclear proton is converted into a neutron and a neutrino is emitted. The reaction is

$$p + e^- \rightarrow n + v.$$

Usually the captured electron comes from the K-shell, and an X-ray photon is emitted.

Electron capture occurs more often than positron emission in heavy nuclei because in them the electron orbits are much nearer.

Because the absorption (capture) of an electron by a nucleus is equivalent to the emission of a positron from it, the electron capture reaction ($p + e^- \rightarrow n + \nu$) is essentially the same as the positron emission reaction ($p \rightarrow n + e^+ + \nu$).

Experimental Investigation of β-ray Energy Spectra : The β-rays emitted from radioactive nuclei consist of electrons with varying high velocities (energies). Their energy variation can be studied by a β-ray spectrometer. The radioactive material (β-ray source) is placed in vacuum upon a fine wire at S (Fig.). The emitted electrons pass through fairly wide slits and are bent by means of a magnetic field B so that they are received by a Geiger counter placed in a fixed position.

The magnetic field B is directed upwards perpendicular to the plane of paper. An electron (mass m, charge e) emitted with a velocity v and entering the field B perpendicularly, describes a circular path in the field. The magnetic force on the electron, evB, supplies the centripetal force mv^2/r, where r is the radius of the path. Thus

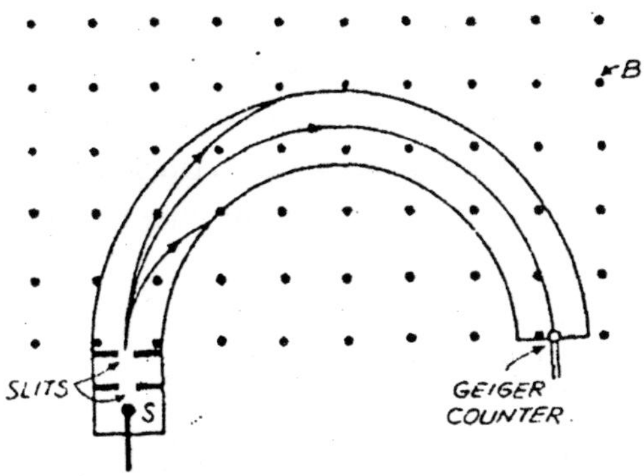

$$evB = mv^2/r$$

or
$$r = \frac{mv}{eB}$$

The position of the counter is fixed. Thus r is fixed.

Therefore, for a given value of B, only those, electrons are recorded by the counter whose momentum is given by

$$p = mv = eBr.$$

The magnetic field B is varied and the number of electrons reaching the counter per unit time is obtained for different values of B. Since each value of B corresponds to a different value of p, the numbers of electrons corresponding to different momenta are obtained. The observed momenta p can be converted into corresponding kinetic energies K from the relativist formula

$$K = \sqrt{\left(m_0^2 c^4 + p^2 c^2\right)} - m_0 c^2$$

where c is the speed of light. A curve between the relative number of electrons and the corresponding kinetic energy is then plotted (Fig.).

Characteristic Features of β-ray Energy Spectra :The β-particles emitted from a given radioactive nucleus have a *continuous* distribution of kinetic energies from 0 to a maximum value $K_{max.}$ The value K_{max}, called the 'end-point energy', is a characteristic of the emitter. It is 1.17 MeV for RaE (Fig.). Most of the β-particles have energies considerably less than 1.17 MeV, and the average energy per particle is about 039 MeV, only about one-third the end-point energy.

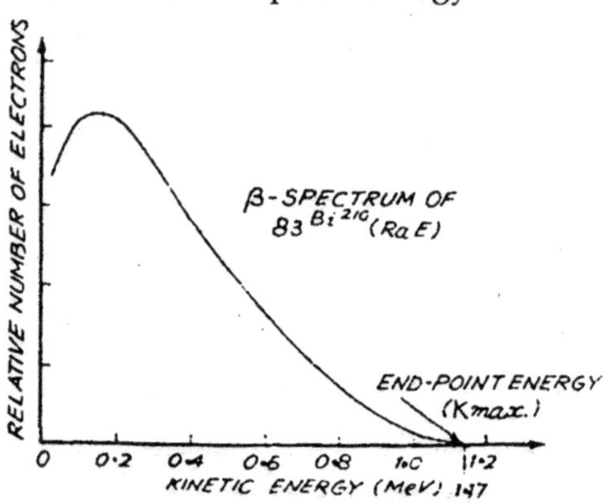

Thus the process of β-disintegration differs from α-disintegration in two important respects. Firstly, the α-particles are already present in the initial nucleus but the β-particles are *not* present in the initial nucleus and are created at the time of emission. Secondly, the energy spectrum of the β-particles is continuous, and not discrete as of α-particles.

Difficulties with β-ray Continuum : In experiments with β-decay the conservation of energy, linear momentum and angular momentum were all appeared to be violated. Let us first consider energy. A parent nucleus in a definite state of energy emits a β-particle and leaves a daughter nucleus which is also in a definite state of energy. Thus the β-particle must emit with a *fixed* energy, equal to the difference between the energies of the parent and the daughter nuclei. But, in practice, the energy spectrum of β-particles is continuous, *i.e.* a β-particle can have *any* kinetic energy between zero and K_{max}.

Meitner argued that all β-particles start from the parent nucleus with same kinetic energy K_{max} but suffer *varying* energy-loses by collision with the atomic electrons surrounding the nucleus. Hence they come out with continuously varying energies. Ellis and Wooster in 1927, performed an experiment to verify this hypothesis. They placed a β-emitting source (RaE) in a thick-walled calorimeter designed to absorb all of the emitted β-particles and measured the total heat (energy) produced by a known number of disintegrations. The heat produced divided by the number of disintegrations gave the average energy per disintegration, $K_{average}$. This was found to be 0.35 MeV, which fairly agreed with the average energy computed from the distribution curve, *but was much less than* K_{max}. Thus Meitner's hypothesis proved to be incorrect. Since the balance of energy, $K_{max}— K_{avarage}$ was unaccountable in β-disintegration, the conservation of energy appeared to be violated.

Linear and angular momenta were also found not to be

conserved in β-decay. If in a single disintegration the directions of the emitted β-particle and of the (recoiling) daughter nucleus are observed (by recording their ionisation tracks), they are not exactly opposite as required for the conservation of linear momentum.

The non-conservation of angular momentum arises because the intrinsic spins of the electron, proton and neutron are all $\frac{1}{2}$. In β-decay a neutron is converted into a proton with the emission of an electron :

$$n \rightarrow p + e^{-}.$$

After the decay, the proton and the electron spins can be parallel (total spin = 1) or anti-parallel (total spin = 0); but in no case the total spin can be $\frac{1}{2}$ (the spin of the original neutron). Thus the spin (and hence angular momentum) is not conserved in the above reaction.

Neutrino Hypothesis of β-Disintegration : In 1930, Pauli suggested that if a particle having zero charge, zero rest mass, and spin $\frac{1}{2}$ is supposed to be emitted together with the electron, then the energy, momentum and angular momentum would all be conserved. Fermi, in 1934, named this particle as 'neutrino' and developed a complete theory of β-disintegration. According to this theory, when a neutron (n) is converted into a proton (p); an electron (e^{-}) and a neutrino (v) are emitted. It was later on found that there are two kinds of neutrino, the neutrino itself (symbol $\bar{v}$) and the antineutrino (symbol $\frac{1}{2}$). In β-disintegration it is the anti-neutrino that is emitted. Thus the basic equation for β-disintegration is

$$n \rightarrow p + e^{-} + \bar{v}.$$

This equation removes the various difficulties encountered in β-disintegration in the following way :

(i) The neutrino (in fact antineutrino) carries no charge. This maintains conservation of charge in the β-disintegration,

(ii) The neutrino has a zero rest mass, and hence a zero rest mass energy. Therefore, in a β-disintegration, the maximum energy which can be carried off by an electron is equal to the energy equivalent of the mass-difference between the parent and the daughter nuclei. If this is E_{max} , then

$$E_{max} = m_0C^2 + K_{max},$$

where m_0c^2 is the rest energy of the electron and K_{max} is the end-point (kinetic) energy of the β-ray spectrum. The energy E_{max} is distributed among the electron, the neutrino, and the recoiling daughter nucleus in a continuous range of different ways. Since the daughter nucleus carries away negligible kinetic energy, the neutrino carries off the difference between K_{max} and the actual kinetic energy of the electron for the particular disintegration. Thus the energy remains conserved in the process.

(iii) The neutrino has a momentum exactly balancing the sum of the momenta of the electron and the recoiling daughter nucleus. Thus momentum remains conserved.

(iv) Like electron, the neutrino has a spin $\frac{1}{2}$. This leads to the conservation of angular momentum because the spins of the three decay particles can combine to give $\frac{1}{2}$.

(v) Lacking mass and charge, the neutrino can pass unhindered through vast amounts of matter. If this were not so, the neutrinos would have been stopped in the calorimeter experiment and their energy would have been absorbed by the calorimeter, giving a temperature rise corresponding to K_{max}.

Line Spectrum of β -rays

Internal Conversion : For certain β-emitters the continuous distribution curve has a number of distinct peaks superimposed upon it (Fig.), which are the characteristics of the emitter. This means that some β-ray sources emit, in addition to the continuous spectrum, a number of β-particles of discrete energies. They are said to constitute a line spectrum of β-rays.

The line spectrum of β-rays arises due to extra nuclear electrons ejected from the atom by a process called 'internal conversion'. A nucleus in an excited state may return to a lower energy state in two ways : (i) by emitting a γ-ray photon of energy hv equal to the difference between (Fig.) the energies of the two nuclear states, (ii) by giving up the energy hv to an electron in the *K*-, *L*-......shell of the same atom. In the second case (internal conversion) the electron is ejected from the shell with a discrete kinetic energy $hv—E_k$, $hv—Ei$,......, where E_k , E_L,....... are the binding energies of, the electron in the K-, L-,....... shells respectively. (The vacancy created in the shell is filled by electrons in higher shells cascading down with the emission of X-rays characteristic of the atom.) Thus the process of internal conversion, which gives rise to β-ray line spectrum, is a *direct* transfer of excitation energy from the nucleus to one of the surrounding electrons. Since in the internal conversion the electrons do *not* come from the nucleus, the not a true form of β-disintegration.

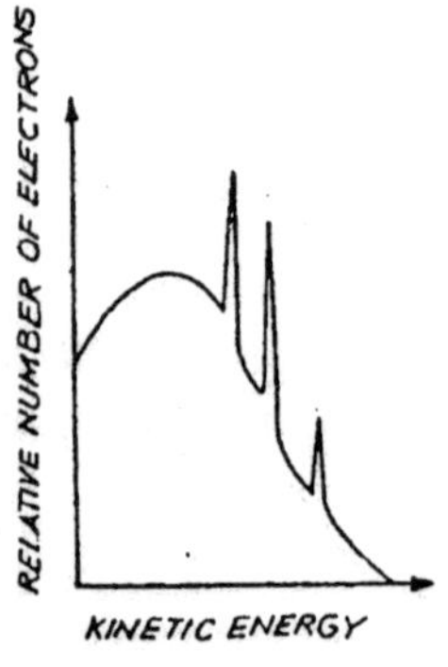

Interaction of γ-rays with Matter : The γ-rays are electromagnetic radiation of very short wavelengths (≈ 0.004 Å to 0.4 A). They have no electric charge, and so they cannot be deflected by magnetic or electric fields. Consequently, direct measurements of their energies (or wavelengths) with a magnetic spectrometer are not possible. The absorption of γ-rays by matter is also different from that of charged particles such as α- and β-rays. Charged particles lose their energy by inelastic collisions so that they slow down and finally come' to rest and absorbed at the end of their range. On the other hand, when a beam of γ-ray photons passes through matter, the intensity of the beam (number of photons) decreases exponentially according to the law

$$I = I_0 e^{-\mu x}$$

where I_0 is the initial intensity of the beam, μ is the absorption coefficient of the substance and x is the thickness of the absorber. Thus γ-rays have no definite range, as do α- and β-rays.

Three separate processes are responsible for the decrease in intensity (absorption) of γ-rays. They are photoelectric absorption, Compton scattering and pair production.

(i) *Photoelectric Absorption :* In this process *all* the energy of a γ-ray photon is transferred to a bound electron, and the γ-ray photon ceases to exist. The ejected electron may either escape from the absorber or reabsorbed due to collisions. At low photon energies (5.0 keV for aluminium and 500 keV for lead), the photoelectric effect is chiefly responsible for the γ-ray absorption.

(ii) *Compton Scattering :* At energies in the neighbourhood of 1 MeV, Compton scattering becomes the chief cause of removal of photons from the γ-ray beam. In this process the γ-ray photon is scattered by one of the atomic electrons which is separated from its atom. The scattered photon moves with reduced energy in a direction different from the original direction and is thus removed from the incident beam.

(iii) *Pair Production* : At high enough energies, both the photoelectric absorption and Compton scattering become unimportant compared with pair production. In the latter process, a γ-ray photon, in passing close to an atomic nucleus in the absorbing matter, disappears and an electron and a positron are created :

$$\underset{(\gamma\text{-photon})}{\gamma} \rightarrow \underset{(\text{electron})}{e^-} + \underset{(\text{positron})}{e^+}$$

The electron and the positron form a pair of particles and hence the process is called 'pair production. The conservation of charge is obvious from the above equation. The rest mass m_0, and hence the rest mass energy m_0c^2, of the positron is the same as that of the electron *i.e.* 0.51 MeV. The energy of the γ-ray photon, $h\nu$, must be atleast $2 \times 0.51 = 1.02$ MeV for pair production to be possible, because this amount of energy is needed to supply the rest energy of the two particles. If $h\nu$ is greater than 1.02 MeV, the balance of energy appears as kinetic energy of the particles (neglecting the small recoil energy of the nucleus).

Measurement of the Wavelengths of γ-rays : There are two chief methods for measuring γ-ray wavelengths.

(i) *By a Crystal γ-ray Spectrometer :* The method is suitable for measuring the wavelengths of the longer γ-rays. A quartz diffracting crystal C (Fig.) is bent and clamped so that the diffracting planes meet, when extended, in a line at A, normal to the plane of the figure. The radius of curvature of the crystal is then equal to the diameter of the focussing circle whose centre is O. Suppose the source of γ-rays is at a point S on the focussing circle such that the Bragg condition ($2d \sin \theta = n\lambda$) is satisfied. Then a diffracted beam enters the detector

D(a scintillation counter) as if it came from the virtual source at S'. For each different wavelength there is a particular position of the source on the focussing circle for which a strong diffracted beam is obtained.

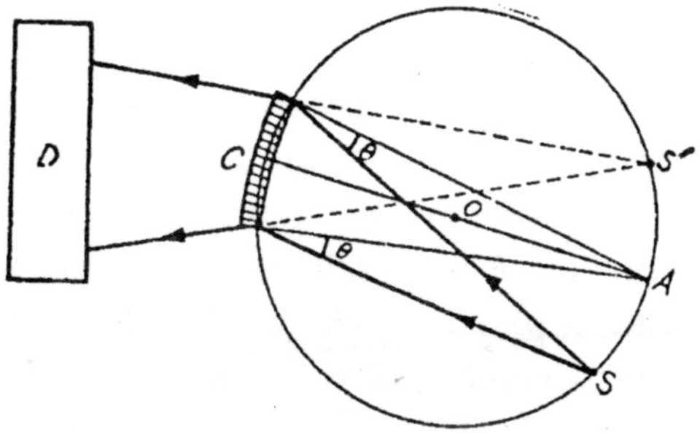

In order to perform the experiment, the source position on the circle is varied, and the corresponding counting rates of the detector are determined. A graph is plotted between the counting rate and the angle θ (position of the source). This graph is found to have sharp peaks which correspond to strong diffracted beams. The angle θ corresponding to a peak is the Bragg's angle. Thus, knowing the crystal lattice spacing d_s the wavelength A of the γ-radiation in a known order can be calculated by using Bragg's equation.

This method has two disadvantages. Firstly, the measurements become more difficult and less precise as the energy of the γ-rays increases and the wavelength decreases, and so the method can be used upto 1 MeV only. Secondly, the method requires highly active γ-ray source.

(ii) *By a Magnetic Spectrograph :* The wavelengths of γ-rays of moderate energy can be determined by an indirect method using a magnetic spectrograph. In this method we obtain a magnetic spectrum of the secondary electrons which are produced by photoelectric absorption or by Compton scattering of γ-rays in matter.

A simple arrangement, used by Ellis, is shown in Fig. The source of γ-rays, such as radium-*B*, is kept in a small thin-walled tube *S* wrapped round with a foil of lead, platinum or tungsten and placed in vacuum. The foil is thick enough to stop all the primary electrons (β-rays) from the radioactive source itself but not the γ-rays. The γ-rays passing out through the wall of the tube eject (secondary) electrons of different energies from the metallic foil due to photoelectric absorption. These electrons pass through a slit *O* and are bent by means of a magnetic field *B* perpendicular to the plane of paper so that they reach a photographic plate *P*. An electron (mass m, charge e) which comes out from the foil with a velocity v describes in the magnetic field a circular path of radius r such that the magnetic force on the electron, evB, supplies the necessary centripetal force mv^2/r. Thus

$$evB = mv^2/r.$$

The momentum of the electron is therefore given by

$$p = mv = eBr. \qquad \text{... } (i)$$

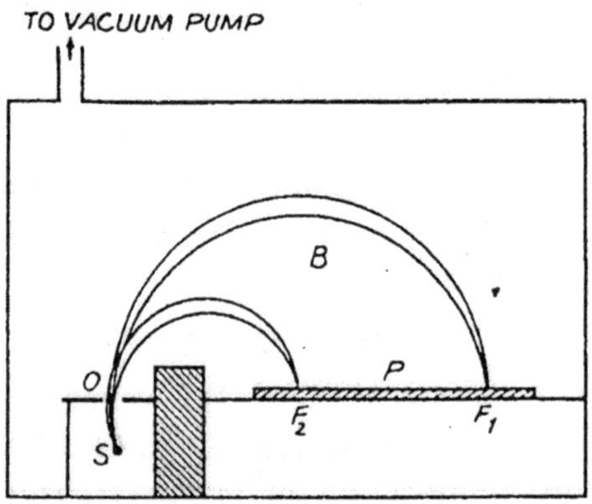

Thus those electrons which have the same momentum have circular paths of the same radii and come to a line-focus at the photographic plate. Hence upon developing the plate, a number of focal lines F_1 F_2, ... are obtained which correspond to various electron momenta (or energies). The various momenta p. can be determined by knowing the magnetic field B and the radii r of the various electron paths. For example, the radius of the path corresponding to F_1 is given by

$$r = \frac{1}{2}(SF_1) = \frac{1}{2}\sqrt{(OS^2 + OF_1^2)}$$

The momenta p as determined from eq. (i), can be converted into corresponding kinetic energies K from the relativistic formula

$$K = \sqrt{(m_0^2c^4 + p^2c^2)} - m_0c^2$$

The electrons ejected from the atoms of the metallic foil must come from one or other of the shells $K, L, M,$....... Suppose, for example, that an electron is ejected from a K-shell. For this, an amount of energy Wk is required which is supplied by the γ-ray incident upon the foil. If hv is the energy of the incident γ-rays, we have

$$hv = W_k + K;$$

where K is the kinetic energy of the ejected electron. The value of W_k for the given atom is known from X-ray studies, whereas K has been obtained from the magnetic spectrograph. Hence hv, and so v (or A), the frequency (or wavelength) of the γ-rays can be determined.

Nuclear Energy Levels—Origin of γ-rays : The discovery of the fine structure of α-rays, and also the experimental fact that γ-rays form a *line* spectrum, gave rise to the idea of nuclear energy levels. It is supposed that there are a number of discrete energy levels in the nucleus, and that a nucleus is normally in its lowest energy state, but it may also exist for short times in an excited state. The transition of the nucleus from an excited state to the lower states gives rise to γ-ray lines.

An important question which arises is how does the nucleus come in an excited state from which it can undergo transitions to lower levels ? An answer is furnished by the fact that in natural radioactivity, γ-rays are emitted only by nuclei which also emit α- or β-rays. Measurements of the α- or β-ray energies and the γ-ray energies show that the γ-rays are emitted by the *daughter* nucleus produced by the emission of an α- or β-particle from the parent nucleus.

Thus it can be supposed that the emission of an α- or β-particle sometimes leaves the daughter nucleus, not in the normal state, but in an excited state of higher energy. The excited daughter nucleus then passes to the normal state with the emission of γ-rays. If it returns to the normal state in a single transition, a single γ-photon is emitted. If it returns by a series of transitions, a series of γ-ray photons is emitted. Measurements of these γ-ray energies help in locating the energy levels of the nucleus.

As an example, radium emits α-particle to become radon in accordance with the transformation

$$_{88}Ra^{226} \rightarrow {}_{86}Rn^{222} + {}_{2}He^{4}$$

(α-particle)

The emitted α-particles fall into two *sharp* energy groups, one having an energy of 4.80 MeV and the other 4.61 MeV. Accompanying them γ-rays of 0.19 MeV energy are also detected (Fig.), The explanation is that, when radium emits an α-particle of 4.80 MeV energy, the daughter nucleus of radon is formed in its ground state. On the other hand, when radium emits an α-particle of 4.61 MeV energy, then the radon nucleus is left in an excited state, from which it passes to the ground state by emitting a γ-ray of energy 4.80 – 4.61 = 0.19 MeV. Similarly, when thorium C decays to thorium C″, five groups of α-particles are emitted, and the existence of four nuclear excited states of thorium C above the ground state is established.

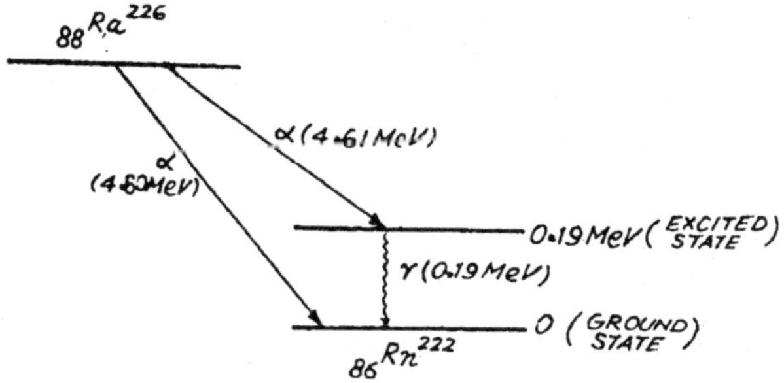

Like α-emission, β-emission also gives information about the energy levels of the daughter nuclei. For example, $_{12}Mg^{27}$ emits β-particle to become $_{13}Al^{27}$:

$$_{12}Mg^{27} \rightarrow {}_{13}Al^{27} + {}_{-1}e^{0}$$

(β-particle)

The emitted β-particles fall into two energy groups, one having an end-point (maximum) energy of 1.78 MeV and the other 1.59 MeV. (In each group the energies are distributed from zero to a maximum). Accompanying them γ-rays of energies 0.834 MeV, 1.015 MeV and 0.181 MeV are also detected. The decay scheme consistent with all of these data is shown in Fig.

The $_{12}Mg^{27}$ nucleus, after emitting the β-particle, leaves the daughter $_{13}Al^{27}$ nucleus in either of its two excited states from which it proceeds to the ground state by γ-ray emission.

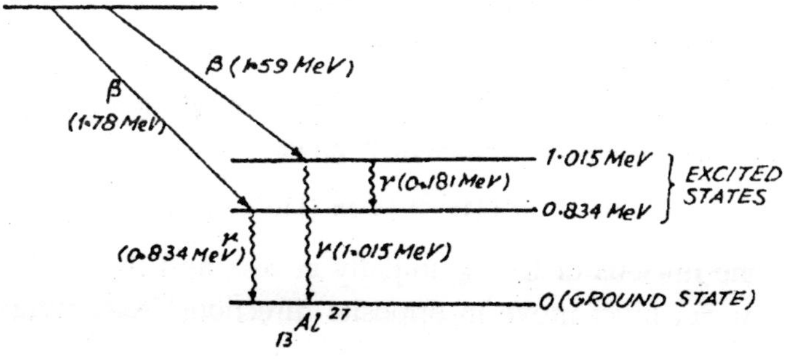

PROBLEMS

1. *Find the kinetic energy of the α-particle emitted in the decay of $_{92}U^{232}$, assuming this atom to be at rest. The kinetic masses of U^{232}, Th^{228} and He^4 are 232.037168, 228.028750 and 4.002603 u respectively. Given : 1 u = 1.66 × 10^{-27} kg and 1 u × c^2 = 931.5 MeV.*

Solution : The decay equation is

$$_{92}U^{232} \rightarrow {}_{90}Th^{228} + {}_2He^4$$

The Q-value (energy) for this decay is given by

$$\begin{aligned} Q &= [m\,(U^{232}) - m\,(Th^{238}) - m\,(He^4)]\,c^2 \\ &= (232.037168\text{ u} - 228.028750\text{ u} - 4.002603\text{ u})\,c^2 \\ &= 0.005815\,u \times c^2. \end{aligned}$$

Now,

$$1\,u \times c^2 = 931.5\text{ MeV}.$$

$$\therefore \quad Q = 0.005815 \times 931.5 = 5.42\text{ MeV}.$$

Since Q-value is positive, the energy is *liberated* and so the decay occurs.

The liberated energy of 5.42 MeV is distributed between the α-particle and recoiling daughter nucleus of $_{90}Th^{228}$, that is,

$$\frac{1}{2} m\,(Th^{228})\,vR^2 + \frac{1}{2}\,m\,(He^4)\,v_a{}^2 = 5.42\text{ MeV}, \qquad \text{... } (i)$$

where v_R and v_a are the velocities of the recoiling nucleus and the α-particle respectively.

The actual distribution can be calculated by linear momentum conservation, which gives

$$0 = m\,(Th^{228})\,v_R - m\,(He^4)\,v_a\,.$$

The nucleus of U^{232} is initially at rest, and the Th^{228} and He^4 nuclei must move in opposite directions. Assuming the ratio of Masses approximately equal to the ratio of mass numbers, we can write from the last equation.

$$v_R = \frac{4}{228} v_\alpha.$$

Substituting this value of v_R in eq. (*i*), and writing it in *SI* units, we have

$$\frac{1}{2} \times (228 \times 1.66 \times 10^{-27} \text{ kg}) \times \left(\frac{4}{228} v_\alpha\right)^2 + \frac{1}{2} \times (4 \times 1.66 \times 10^{-27} \text{ kg}) \times v_\alpha^2$$

$$= 5.42 \times 1.6 \times 10^{-13} \text{ joule.}$$

Solving, we get

$$v_\alpha = 1.60 \times 10^7 \text{ m/s.}$$

$$\therefore \quad K_\alpha = \frac{1}{2} m(\text{He}^4) v_\alpha^2$$

$$= \frac{1}{2} \times (4 \times 1.66 \times 10^{-27} \text{ kg}) \times (1.60 \times 10^7 \text{ m/s})^2$$

$$= 8.50 \times 10^{-13} \text{ joule.}$$

$$= \frac{8.50 \times 10^{-13}}{1.6 \times 10^{-13}} = 5.31 \text{ MeV.}$$

2. ***$_{10}Ne^{23}$ decays to $_{11}Na^{23}$ by beta emission. What is the maximum kinetic energy of the emitted electrons ? The atomic masses of Ne^{23} and Na^{23} are 22.994466 u and 22.989770 u respectively.***

Solution: The decay equation is

$$_{10}\text{Ne}^{23} \rightarrow {}_{11}\text{Na}^{23} + {}_{-1}\beta^\circ + \bar{v}.$$

The *Q*-value of this decay is given by (ignoring electron mass)

$$Q = [m(\text{Ne}^{23}) - m(\text{Na}^{23})]\, c^2$$

$$= (22.994466 \text{ u} - 22.989770 \text{ u})\, c^2$$

$$= 0.004696 \text{ u} \times c^2$$

$$= 0.004696 \times 931.5 = 4.37 \text{ MeV.}$$

Except for a small correction for the kinetic energy of the recoiling Na^{23} nucleus, the *maximum* kinetic energy of the electrons is just equal to this.

3. Compute the minimum energy of a γ-ray photon which may produce an electron-positron pair. The rest mass of electron is 91 × 10^{−31} kg and the speed of light is 30 ×10⁸ m/s.

Or

Why does a photon of wavelength 1 Å cannot produce an electron-positron pair ?

Solution: When an energetic γ-rays photon falls on a heavy substance, it is absorbed by some nucleus of the substance and an electron and a positron are produced. This phenomenon is called 'pair-production, and may be represented by the following equation :

$$\underset{(\gamma\text{-photon})}{h\upsilon} = \underset{(\text{positron})}{{}_{+1}\beta^0} + \underset{(\text{electron})}{{}_{-1}\beta^0}$$

According to Einstein's mass-energy relation, every body in the state of rest has some energy, called its rest-mass energy. If the rest-mass of the body be m_0, then its rest-mass energy is

$$E_0 = m_0c^2.$$

The rest-mass of each of the electron and the positron is 9.1×10^{-31} kg. So, rest-mass energy of each of them is

$$E_0 = m_0c^2$$

$$= (9.1 \times 10^{-31} \text{ kg}) \times (3.0 \times 10^8 \text{ m/s})^2$$

$$= 8.2 \times 10^{-14} \text{ joule}$$

$$= \frac{8.2 \times 10^{-14}}{1.6 \times 10^{-13}} = 0.51 \text{ MeV.}$$

[because 1 MeV = 1.6×10^{-13} joule]

Hence, for pair-production, it is essential that the energy of γ-photon must be at least $2 \times 0.51 = 1.02$ MeV.

The wavelength corresponding to this minimum photon-energy is

$$\lambda = \frac{hc}{E} = \frac{\left(6.62 \times 10^{-34} \text{ joule-sec}\right) \times \left(3.0 \times 10^{8} \text{ meter/sec}\right)}{\left(1.02 \times 1.6 \times 10^{-13} \text{ joule}\right)}$$

$$= 1.2 \times 10^{-12} \text{ meter} = 0.012 \text{ Å}$$

This is the *maximum* wavelength of the photon to produce electron-positron pair. Hence a 1.Å photon cannot produce the pair.

5

Atomic Mode

α-particle Scattering

In 1906, Rutherford observed that when a sharp beam of α-particles falls upon a photographic plate in vacuum, a sharp image is obtained. If, however, a thin foil of metal is placed in the path of the beam, the image becomes diffuse. This is due to the scattering of α-particles by the atoms of the foil.

In 1909, Geiger and Marsden made a series of measurements on the scattering of α-particles. They placed a thin metal gold foil in the path of a ray of α-particles from a radioactive substance (Fig. below). A fluorescent screen *S*, backed by a microscope *M*, was placed behind the foil. The particles, while passing through the foil deflected (scattered) through a wide range of angles.

The number of particles scattered in various scattered in various directions were obtained by counting the scintillations they made on the screen *S*. The screen was always maintained perpendicular to the incident ray. It was found that *although most of the particles scattered through angles of the order of 1° or less, out a small number, say about 1 in every 10,000, scattered through 90° or even 180°.* This *large-angle* scattering of α-particles could not be at all explained from the-then prevalent ideas about atom.

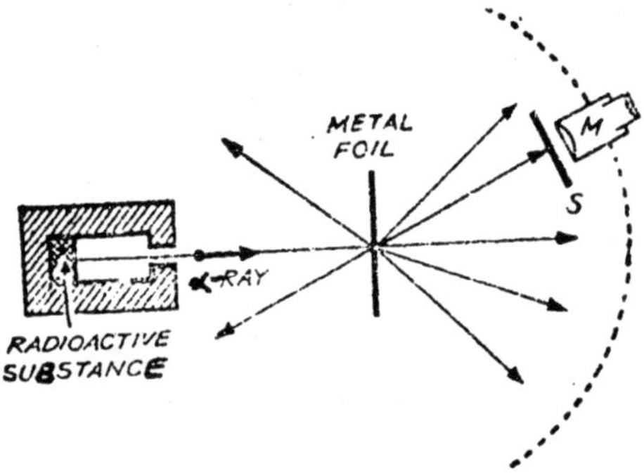

In 1911, Rutherford explained, successfully the large-angle scattering by proposing a structure of the atom. He made the following assumptions :

(i) The positive charge of the atom is concentrated in an extremely small volume, known as nucleus', in the atom negative charges (electrons) revolve around the nucleus in planetary orbits. Thus most of the atom consists of empty space. This is why a great majority of α-particles pass the metal foil almost undeviated.

(ii) The whole mass of the atom (excluding the electrons) is contained in the nucleus along with the positive charge.

(iii) The scattering of α-particles by atoms takes place due to an encounter between the a-particle and the *nucleus* of the atom. The electrons outside the nucleus, owing to their little mass, cannot appreciably deflect the far more massive α-particles.

(iv) Both the α-particle and the nucleus are small enough

to be considered as *point* mass and *point* (positive) charge; and the electrostatic repulsive force between them is governed by Coulomb's inverse square law,

(v) The nucleus of the atom (of the metal foil used) is so massive compared with the α-particle that it remains at rest during the encounter.

When an α-particle strikes the metal foil, it can penetrate the outer electron-cloud of an atom and approach closely the nucleus. It then moves under the action of a coulombian force of repulsion, and its path is a *hyperbola* with the nucleus as the external focus.

Rutherford's Scattering Formula

The Rutherford's assumptions regarding α-scattering have been given in the last question.

Let us consider an α-particle moving in a direction PO (Fig.), where *P* is far from a stationary nucleus *N*. As the particle approaches the nucleus, it begins to be repelled more and more and assumes the hyperbolic path *PAP'*. *OP* and *OP'* make equal angles, 6, with the line *NA* where *A* is the apex of the hyperbola. Let *p* be the perpendicular distance from the

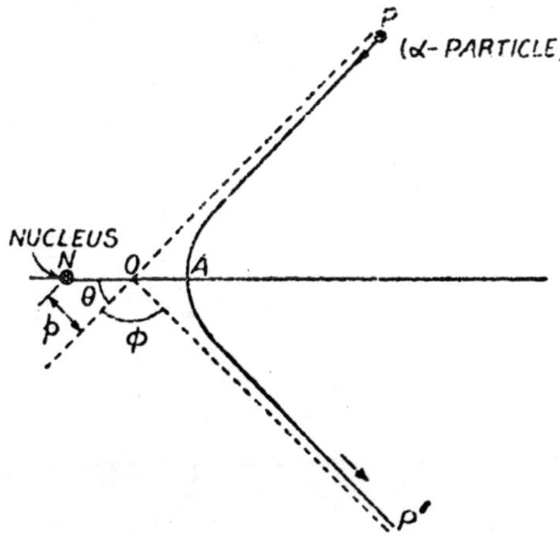

nucleus N to the original direction of the α-particle. v is called the 'impact parameter'. It is the minimum distance to which the α-particle would approach the nucleus if there were no force between them. The angle between the direction of approach and the direction of recede of the α-particle is the 'scattering angle' ϕ. Let us first find a relation between p and ϕ.

Let m_α. be the mass of the α-particle, v its velocity at P, and v' the velocity at A. By conservation law, the initial angular momentum* of the α-particle about the nucleus will be equal to its angular momentum when at A. Therefore

$$m_x \, v \, P = m_\alpha \, v' \, a,$$

where $$a = NA.$$

Thus

$$v = \frac{v\,p}{a} \qquad \text{... } (i)$$

We know that the (positive) charge on the nucleus is Ze, where Z is the atomic number, and that on the α-particle is $2e$, where e is the electronic charge. Therefore, the electrostatic potential energy of the α-particle when at A will be $\frac{1}{4\pi_0^\epsilon} \frac{(Ze)(2e)}{a}$. Now, from the conservation of energy, the initial kinetic energy of the α-particle is equal to the sum of the kinetic and potential energies at A, so that

$$\frac{1}{2} m_\alpha \, v^2 = \frac{1}{2} m_\alpha \, v^2 + \frac{1}{4\pi_0^\epsilon} \frac{2Ze^2}{a}$$

or $$v'^2 = v^2 - \frac{1}{4\pi_0^\epsilon} \frac{4Ze^2}{m_\alpha \, a}$$

or $$v'^2 = v^2 \left(1 - \frac{1}{4\pi_0^\epsilon} \frac{4Ze^2}{m_\alpha \, c^2 \, a}\right)$$

Let us introduce E_k $(=\frac{1}{2} m_\alpha v^2)$, *the initial* kinetic energy of the α-particle, instead of its mass m_z and velocity v. Then we have

$$v'^2 = v^2 \left(1 - \frac{1}{4\pi\epsilon_0} \frac{4Ze^2}{E_k\, a}\right)$$

or
$$v'^2 = v^2 \left(1 - \frac{b}{a}\right),$$

where
$$b = \left(\frac{1}{4\pi\epsilon_0} \frac{2Ze^2}{E_k}\right).$$

Now, putting the value of v' from eq. (*i*), we get

$$\frac{v^2\, p^2}{a^2} = v^2 \left(1 - \frac{b}{a}\right),$$

or
$$p^2 = a^2 - ba.$$

N is the external focus and A the apex of the hyperbola. It can be shown from the coordinate geometry that

$$a = NA = p \cot(\theta/2),$$

where θ is the angle shown. Therefore the above expression becomes

$$p^2 = p^2 \cot^2(\theta/2) - bp \cot(\theta/2).$$

$$\therefore \quad b = p\, \frac{\cot^2(\theta/2) - 1}{\cot(\theta/2)}$$

$$= p\, \{\cot(\theta/2) - \tan(\theta/2)\}$$

$$= p \left\{\frac{\cos(\theta/2)}{\sin(\theta/2)} - \frac{\sin(\theta/2)}{\cos(\theta/2)}\right\}$$

$$= p \left\{\frac{\cos^2(\theta/2) - \sin^2(\theta/2)}{\sin(\theta/2)\cos(\theta/2)}\right\}$$

$$= 2p \left\{ \frac{\cos\theta}{\sin\theta} \right\} = 2p \cot\theta.$$

$$\therefore \quad \cot\theta = \frac{b}{2p}. \qquad \text{...}(ii)$$

From the fig., the scattering angle ϕ is equal to π—2θ, so that

$$\frac{\phi}{2} = \frac{\pi}{2} - \theta$$

or
$$\cot\frac{\phi}{2} = \cot\left(\frac{\pi}{2} - \theta\right) = \tan\theta.$$

But tan
$$\theta = \frac{2p}{b} \text{ from eq. } (ii).$$

$$\therefore \quad \cot\frac{\phi}{2} = \frac{2p}{b}. \qquad \text{...}(iii)$$

This shows that the scattering angle ϕ decreases *rapidly* as the impact parameter p increases. It means that *only an α-particle passing extremely close to the nucleus suffers large-angle scattering.*

Eq. (*iii*) cannot be verified experimentally because we cannot measure the impact parameter p for an observed angle ϕ. We therefore calculate the number of α-particles scattered through an angle ϕ. We see from eq. (iii) that all α-particles, such as 1, 2, 3 in the below–given figure , which approach the nucleus with an impact parameter less than p will be scattered through an angle greater than ϕ. This means that an α-particle that is initially directed anywhere within the area πp^2 around the nucleus will be scattered through an angle greater than ϕ. Let t be the thickness of the scattering foil and n the number of atoms per unit volume. Then, it can be shown from the theory of probability that the fraction f of incident α-particles scattered through angle greater than ϕ is

$$f = \pi p^2 nt.$$

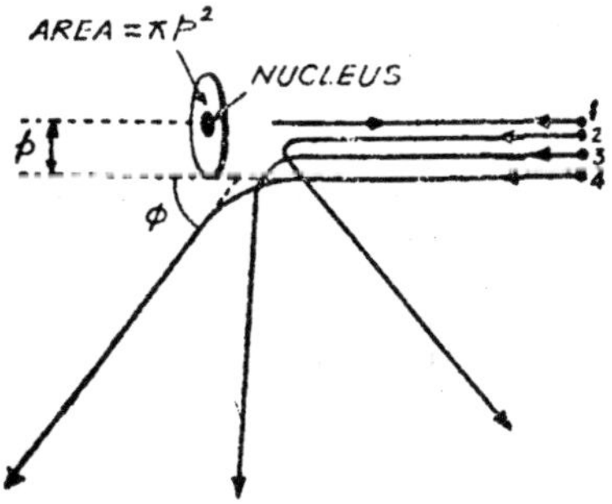

We assume that the foil is sufficiently thin so that an α-particle receives the entire deflection from a single encounter with a nucleus. Then we can substitute the value of p $\left(=\frac{b}{2}\cot\frac{\phi}{2}\right)$ from eq. (*iii*). Thus

$$f = \frac{1}{4}\,\pi b^2 nt \cot^2\frac{\phi}{2}.$$

Therefore, the fraction of incident α-particles scattered between ϕ and $\phi + d\phi$ is

$$df = \frac{1}{4}\,\pi b^2 nt \cot\left(\frac{\phi}{2}\right)\text{cosec}2\left(\frac{\phi}{2}\right)\,d\phi. \quad \text{... } (iv)$$

The minus sign indicates that f decreases with increasing ϕ.

In the experiment, a fluorescent screen was placed at a distance r from the foil, and the scattered α-particles were counted by means of the scintillations they produced on the screen. Those particles which scattered between ϕ and $\phi + d\phi$ reach a spherical zone of radius $r \sin\phi$ and width $r\,d\phi$ (Fig.). Thus the area dS of the screen struck by these particles is

$$\begin{aligned} DS &= (2\pi\, r \sin\phi)\,(r\,d\phi) \\ &= 2\pi\, r^2 \sin\phi\; d\phi \\ &= 4\pi\, r^2 \sin(\phi/2)\cos(\phi/2)\,d\phi. \quad \text{... } (v) \end{aligned}$$

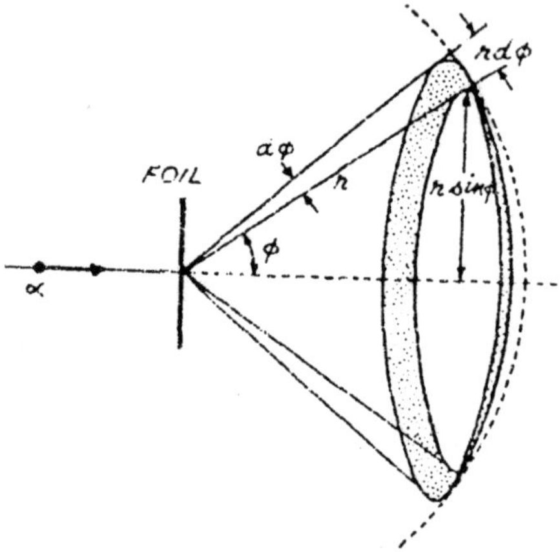

If N_0 be the total number of α-particles striking the foil during the experiment, the number scattered between ϕ and $\phi + d\phi$ is $N_0\, df$. Therefore, the number $N(\phi)$ counted per unit area on the screen placed at angle ϕ, which is the quantity actually measured, is given by

$$N(\phi) = \frac{N_0 |df|}{dS}.$$

Putting the values of df and dS from eq. (iv) and (v), we get

$$N(\phi) = \frac{N_0 (1/4) \pi b^2\, nt \cot(\phi/2) \operatorname{cosec}^2(\phi/2)\, d\phi}{4\pi r^2 \sin(\phi/2) \cos(\phi/2)\, d\phi}$$

$$= \frac{N_0 b^2\, nt}{16 r^2} \frac{1}{\sin^4(\phi/2)}$$

Substituting the value of $b \left(= \frac{1}{4\pi\epsilon_0} \frac{2Ze^2}{E_k} \right)$, we obtain

$$\boxed{N(\phi) = \frac{N_0\, ntZ^2 e^4}{4r^2 (4\pi\epsilon_0)^2 E_k^2} \frac{1}{\sin^4(\phi/2)}.} \qquad ...(vi)$$

This is Rutherford's scattering formula for the number of α-particles scattered at a particular angle (ϕ).

Significance of Rutherford's Formula

The formula shows that the number of scattered α particles arriving per unit area at the screen should be directly proportional to the thickness t of the foil and to the square of the atomic number Z of the foil atoms ; and it should be inversely proportional to the square of the kinetic energy E_k of the α-particles and to $\sin^4 (\phi/2)$, where ϕ is the scattering angle. All these predictions agreed with the measurements of Geiger and Marsden who performed the experiment by using scattering foils of different thicknesses and of different materials, and the α-particles of different energies. In 1920, Chadwick used the formula to measure correctly the charge Ze on the nucleus for a number of elements, thus giving a further verification of the formula. Hence we conclude that Rutherford's assumption of the existence of a tiny, massive, positive nucleus in the atom is correct. Rutherford is therefore credited with the "discovery" of the nucleus.

Scattering Angle

The angle of scattering ϕ of an α-particle is related to the impact parameter p by

$$\cot \frac{\phi}{2} = \frac{2p}{b},$$

where $b = \frac{1}{4\pi\epsilon_0} \frac{2Ze^2}{E_k}$, E_k being the initial kinetic energy of the particle.

Now, b is same for the two α-particles, because E_k is same. Larger the scattering angle ϕ, smaller will be the impact parameter p. Clearly, the particle scattered through a *larger* angle comes closer to the nucleus.

It is found that the number of α-particles scattered through very small angles is somewhat different from the number predicted by Rutherford's formula. The reason for the same is as follows.

Rutherford's formula fails to agree with the experimental data at very small scattering angles. This is because small ϕ implies a large impact parameter, in which case the full nuclear charge of the target atom in partially screened by its electrons.

Distance of Closest Approach : When α-particles falls upon a thin metal foil, the particles are scattered in various directions by the *nuclei* of the atoms of the foil. The angle of scattering of a particular particle depends upon the particle's impact parameter (the perpendicular distance from the nucleus to the initial direction of motion of the particle). Rutherford showed that if the nucleus be assumed to be a *point mass* and a *point charge* and the electrostatic force between the nucleus and the α-particle be assumed to be governed by Coulomb's inverse-square law, then the angle of scattering is related to the impact parameter p by the equation

$$\boxed{\cot\left(\frac{\phi}{2}\right) = \frac{2p}{b},}$$

where $b\left(=\frac{1}{4\pi\epsilon_0}\frac{2Ze^2}{E_k}\right)$ depends upon the charge of the given nucleus and the kinetic energy of the smaller the incident α particle. This smaller the impact parameter p, greater the scattering angle . When $p = 0$, then ϕ assumes the maximum value of 180°. This means that when the α-particle moves directly (head-on) toward the nucleus, it is scattered through 180°, *i.e., reflected back* along its initial path (Fig.). Clearly, it is under this condition that the particle will have its closest approach to the nucleus.

Let r_0 be the distance of closest approach of the α-particle to the nucleus. The (positive) charge on the nucleus is Ze, and that on the α-particle is $2e$, where e is the electronic charge.

Therefore, the electrostatic potential energy of the particle at the instant of closest approach is $\frac{1}{4\pi\epsilon_0}\frac{(Ze)(2e)}{r_0} = \frac{1}{4\pi\epsilon_0}\frac{2Ze^2}{r_0}$. At this instant the α-particle is momentarily at rest, and the initial kinetic energy E_k is entirely converted into electrostatic potential energy. Hence at this instant

$$E_k = \frac{1}{4\pi\epsilon_0}\frac{2Ze^2}{r_0}$$

$$\boxed{\therefore\ r_0 = \frac{1}{4\pi\epsilon_0}\frac{2Ze^2}{E_K}.}$$

This is the expression for the distance of closest approach r_0 of the α-particle. It shows that for a given nucleus, r_0 depends upon the initial kinetic energy E_k of the α-particle.

When the kinetic energy E_k exceeds a certain value, the distance of closest approach r_0 becomes so small that the nucleus no longer appears as a *point* charge to the α-particle. Then the Coulomb's inverse-square law and hence the Rutherford formula breaks down. Besides this, some short-range *attractive* nuclear forces also begin to operate. Hence under this condition the incident particle is not reflected back by the nucleus ; instead the particle penetrates the nucleus. Thus, *there is an upper limit for the energy of the incident particle beyond which the particle cannot be reflected back by a given nucleus.*

Estimation of Nuclear Size—Upper Limit of Nuclear Radius : The Rutherford scattering formula holds so long the nucleus appears as *a point* to the scattering α-particle *i.e.* so long the size of the nucleus is insignificant compared with the closest distance r_0 to which the α-particle approaches the nucleus. The more energetic the incident a-particle; the smaller is r_0. Hence the distance of the closest approach, r_0, of a particle which has a maximum energy (upto which the Rutherford formula agrees with experiment) is the upper limit for the nuclear radius.

The experimental observations on α-particle scattering are found to agree with formula upto a maximum energy of 7-8 MeV. For α-particles of this energy the distance of closest approach is about 3×10^{-14} meter for a gold nucleus and 2×10^{-14} meter for a silver nucleus. Hence we conclude that the radius of the gold nucleus is less than 3×10^{-14} meter, and that of the silver nucleus is less than 2×10^{-14} meter. In general, the nucleus is well under 1/10,000 the radius of the atom as a whole.

The *actual* nuclear radii have been estimated by performing scattering experiments using highly energetic electrons and neutrons and studying the deviation of the results from the Rutherford formula. It has been found that the nuclear radius R is given by

$$R = R_0 A^{1/3}$$

where A is the mass number of the nucleus and

$$R_0 \approx 1.3 \times 10^{-15} \text{ meter.}$$

Atomic Structure

Matter, despite its appearance of being continuous, possesses a definite particle structure of a microscopic level. The ultimate particles of matter are known as 'atoms'. Today we know that an atom has its own ultimate particles, electrons, protons, and neutrons.

The first information about the structure of atom came from the discovery of the electron in 1897, when it became known that all atoms contain the negatively-charged electrons. Since matter is electrically-neutral, it was inferred that atoms must also contain positively-charged material. Furthermore, an electron was found thousands of times lighter than the entire atom. This suggested that the positive charge of the atom carries nearly the entire mass. The radius of the atom was estimated to be of the order of 10^{-10} meter from kinetic theory considerations. The distribution of electron and positive charge in the atom was not known however.

Thomson's Model of Atom: Thomson, in 1904, suggested that an atom is a *uniform* sphere of positively-charged material, in which electrons are embedded in number such as to balance the positive charge (Fig.). The model was however, soon discarded as it was in complete disagreement with experiments on the *large-angle* scattering of α-particles by matter.

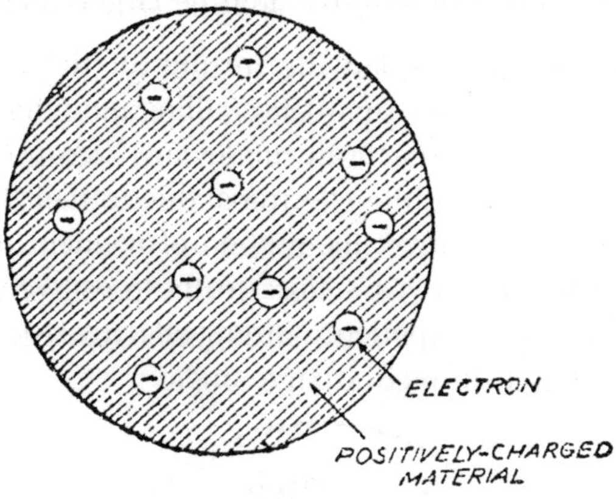

Rutherford's Nuclear Model of Atom: In 1909, Geiger and Marsden made a series of experiments on the scattering of α-particles. They found that α-particles falling on a thin metal foil scattered in various directions. Although most of the particles scattered through angles of the order of 1° or less, but a small number scattered through 90° or even 180°. This led Rutherford to conclude that while most of the (positively-charged) α-particles, while traversing matter, experienced little force, but a few of them experienced *strong* repulsive force as could be exerted by extremely dense concentrations of positive charge. Hence the positive charge of the atom could not be uniformly distributed as pictured by Thomson.

Rutherford suggested that in an atom the entire positive charge and nearly all of its mass are concentrated at the centre of the atom in a small volume, known as the 'nucleus' of the atom (Fig.). The electrons revolve around the nucleus in planetary orbits at distances large compared with the size of the nucleus. The orbital motion was assumed because without it the electrons would fall into the nucleus under the electrostatic attraction and the atom would collapse.

The radius of the nucleus was estimated to be of the order of 10^{-15} meter which is only 1/10,000 of the radius of the atom as a whole. Thus Rutherford's atom is largely hollow.

The model can explain the results of the α-scattering experiment. Most of the α-particles falling upon the metal foil go through the atoms practically undeviated because the atoms are largely hollow and the electrons, due to their little mass, do not appreciably affect the motion of incident α-particles. Those particles which pass close to the (positively-charged) nucleus, however, experience strong electrostatic repulsion and are scattered through a large angle.

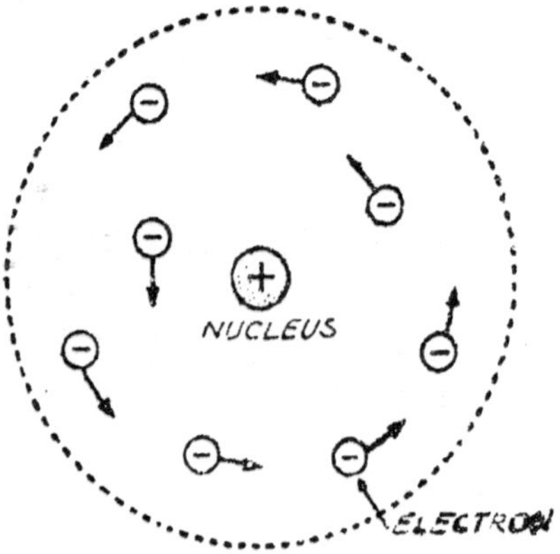

The Rutherford's model was supported by the periodic table of elements also. The α-scattering experiments enabled direct determination of the nuclear charge of the atom involved. It was found that all atoms of any one element had the same unique nuclear charge, and this charge increased regularly from element to element in the periodic table. The nuclear charge for an element was always found to be $+Ze$, where Z determined the position of the element in the periodic table. Z is now known as the 'atomic number' of the element, and gives the number of unit positive charges in a nucleus of that element.

Difficulties in Rutherford's Model (Failure of Classical Physics): The Rutherford's model, in spite of strong experimental support, faced certain difficulties. The revolving electrons are constantly accelerated (centripetal) towards the nucleus. Such electrons, according to electromagnetic theory, would constantly radiate energy in the form of electromagnetic waves. Hence they would rapidly spiral in and fall into the nucleus, and the atom would collapse. (Calculations show that only about 10^{-10} sec would be required for a stable hydrogen atom to collapse). In practice, atom do not collapse.

Furthermore, in Rutherford's model the electrons can revolve in orbits of all possible radii and so they should emit continuous radiation of all frequencies. But the experimental fact is that atoms like hydrogen emit line spectra of radiation of only certain fixed frequencies.

From these difficulties we can only conclude that the classical laws of Physics which are true in macroscopic world do not hold in the microscopic world of the atom.

Bohr's Quantum Model

In 1913, Bohr gave a new and rather bold idea of atomic structure in order to explain the stability of the atom and the emission of sharp spectral lines. He proposed the following two modifications in Rutherford's model which are contrary to the classical theory :

1. The elections can revolve around the nucleus only in certain discrete orbits, and they do not radiate energy while in these permitted orbits. Hence the atom is stable. The permitted orbits are those in which the angular momentum *mvr** of the electron is an integral multiple of $\frac{h}{2\pi}$; *h* being the Planck's constant,

   i.e. $$\text{mvr} = \frac{nh}{2\pi}$$

   where $n = 1, 2, 3, ...$

2. The emission (or absorption) of radiation by the atom takes place when an electron jumps from one permitted orbit to another (Fig.). The radiation is emitted (or absorbed) as a single quantum (photon) whose energy hv is equal to the difference in energy ΔE of the electron in the two orbits involved. Thus

$$hv = \Delta E,$$

where v is the frequency of emitted (or absorbed) radiation. Hence the spectrum of the atom will have certain fixed frequencies.

The main objection against the Bohr's model was that its assumptions were entirely arbitrary, and could only be justified by the experimental results.

The Bohr's model as it is, was found lacking in explaining intensity of spectral lines, fine structure of spectral lines, spectra of complex atoms, etc.

Wave Mechanical Model: de Broglie gave the physical interpretation of Bohr's postulate that only certain orbits are permitted for the electron. He treated the electron (mass *m*, velocity *v*) as a wave of wavelength

$$\lambda = \frac{h}{mv},$$

which forms a stationary wave-pattern along the atomic orbits. He deduced that an electron can revolve around the nucleus indefinitely without radiating energy provided that its orbit contains an integral number of electron wavelengths (Fig.) *i.e.*

$$2\pi r = h\lambda.$$

But $$\lambda = \frac{h}{mv}$$

$\therefore$ $$2\pi r = \frac{h}{mv}$$

or $$mvr = n\frac{h}{2\pi},$$

which is same as the Bohr's postulate.

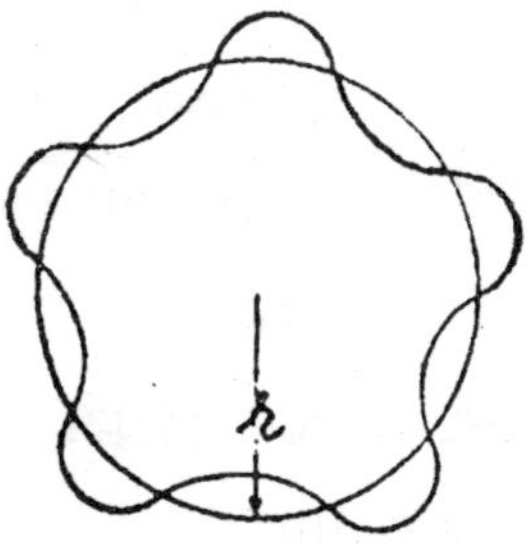

Later on, the application of quantum mechanics to the atom automatically yielded all the results of the Bohr theory and solved many problems which could not be handled by the Bohr theory. The fine structure of spectral lines was explained by considering the relativist variation of electron mass.

PROBLEMS

1. A 5-MeV α-particle approaches a gold (Z= 79) nucleus with an impact parameter of 2.6 × 10^{-15} meter. The angle through which it will be scattered is as follows (e = l.6 × 10^{-19} coulomb and 1 eV = l.6 × 10^{19} joule).

Solution : The angle of scattering ϕ of α-particle is related to the impact parameter p by

$$\cot(\phi/2) = \frac{2p}{b},$$

where $b = \frac{1}{4\pi\epsilon_0}\frac{2Ze^2}{E_k}$, E_k being the initial kinetic energy of the α-particle. Thus

$$\cot(\phi/2) = \frac{2p}{b} = \frac{(4\pi\epsilon_0)\,pE_k}{Ze^2}.$$

Here $p = 2.6 \times 10^{-13}$ meter, $Z = 79$; $e = 1.6 \times 10^{-19}$ coulomb and $E_k = 5$ MeV $= 5 \times 10^6$ eV $= 5 \times 10^6 \times (1.6 \times 10^{-19})$ joule.

Also $\frac{1}{4\pi\epsilon_0} = 9 \times 10^9$ nt-m^2/joul^2.

$$\therefore \qquad \cot(\phi/2) = \frac{(2.6 \times 10^{-13}) \times (5 \times 10^6 \times 1.6 \times 10^{-19})}{(9 \times 10^9) \times 79 \times (1.6 \times 10^{-19})^2}$$

$$= 11.42$$

or $$\phi/2 = \cot^{-1}(11.42) = 5^\circ$$

$$\therefore \qquad \phi = 10^\circ$$

2. A gold foil of density 19.3 gm/cm^3 and atomic weight 197 gm/gm-atom has a thickness of 2.0×10^{-4} cm. It is used to scatter α-particles of kinetic energy 80 MeV. The fraction of the α-particles is scattered at angles greater than 90° is given below For gold Z = 79. (1 MeV=l.6 × 10^{13} J.) Avogadro's number is 6.02 × 10^{23} atoms/gm-atom.

Solution : Let us first calculate the impact parameter p for scattering at 90°. We have

$$\cot\frac{\phi}{2} = \frac{2p}{b},$$

where $b = \dfrac{1}{4\pi\epsilon_0}\dfrac{2Ze^2}{E_k}$. From this,

$$p = \frac{1}{4\pi\epsilon_0}\frac{Ze^2}{E_k}\cot\frac{\phi}{2}$$

$$= (9 \times 10^9\ N\text{-}m^2/C^2)\ \frac{79 \times \left(1.6 \times 10^{-19}\ C\right)^2}{8.0 \times 1.6 \times 10^{-13}\ J}\ \cot 45^{\circ}$$

$$= 1.4 \times 10^{-14}\ m.$$

The cross-section is

$$\pi p^2/ = 3.14 \times (1.4 \times 10^{-14})^2$$

$$= 6.2 \times 10^{-28}\ \text{m}^2/\text{atom}.$$

The number of atoms per unit volume of the foil

$$n = \frac{\text{atoms}}{\text{volume}} = \frac{\text{mass/volume}}{\text{mass/atoms}}$$

$$= \frac{\text{density}}{\text{atomic wt/ Avogadro's number}}$$

$$= \frac{19.3\ \text{mg}/\text{cm}^3}{\left(197\ \text{gm}/\text{gm-atom}\right)/\left(6.02 \times 10^{23}\ \text{atom}/\text{gm-atom}\right)}$$

$$= 5.9 \times 10^{22}\ \text{atoms/cm}^3$$

$$= 5.9 \times 10^{28}\ \text{atoms/m}^3.$$

Now, the fraction of the incident particles scattered through an angle greater than ϕ is given by

$$f_{>\phi} = \pi p^2\, nt,$$

where p is impact parameter for scattering angle ϕ and t is thickness of the foil. Thus

$$f_{>90^{\circ}} = (6.2 \times 10^{-28}\ \text{m}^2/\text{atom})\ (5.9 \times 10^{28}\ \text{atoms/m}^3)\ (2.0 \times 10^{-6}\ m)$$

$$= 7.3 \times 10^{-5}.$$

3. A beam of α-particles of energy 5.5 MeV is incident normally on a gold (Z=79) foil at a rate of 10^4 particles/sec. An α-particle counter with an aperture 4 cm^2 is placed at a distance 0.25 in from the foil. The counts at an angle of 60° is calculated, if the thickness of the foil is 10^{-5} cm. The atomic weight (mass of 1 gm-atom) of gold is 197 and its density is 193 gm/cm^3. The Avogadro's number is 6.023 × 10^{23}.

Solution : The number of α-particles entering the counter per unit area at an angle of φ is given by

$$N(\phi) = \frac{N_0 nt Z^2 e^4}{4r^2\left(4\pi\epsilon_0\right)^2 E_k^2} \frac{1}{\sin^4(\phi/2)}$$

Here, N_0 (particles striking the foil) = 10^4/sec,

n (number of atoms in the foil per unit volume)

$$= \frac{\text{mass / volume}}{\text{mass / no. of atoms}}$$

$$= \frac{\text{density}}{\text{atomic wt / Avogadro's number}}$$

$$= \frac{19.3 \text{ gm / cm}^3}{197 \text{ gm / } 6.023\ \ 10^{23}}$$

$$= 5.9\ 5\ 10^{22}/\text{cm}^3 = 5.9 \times 10^{28}/\text{m}^3,$$

t (thickness of foil) = 10^{-5} cm = 10^{-7} m,

r (distance of counter from the foil) = 0.25 m,

Z = 79, $e = 1.6 \times 10^{-19}$ coul,

E_k = 5.5 MeV=5.5 × 1.6 × 10^{-13}

joule and $\phi = 60°$ so that $\sin \phi/2 = \sin 30° = \frac{1}{2}$. Thus

$$N(\phi) = \frac{10^4 \times \left(5.9 \times 10^{28}\right) \times 10^{-7} \times (79)^2 \times \left(1.6 \times 10^{-19}\right)^4}{4 \times (0.25)^2 \times \left(\frac{1}{9 \times 10^9}\right)^2 \times \left(5.5 \times 1.6 \times 10^{-13}\right)^2} \times \frac{1}{\left(\frac{1}{2}\right)^2}$$

$$= \frac{5.9 \times (79)^2 \times (1.6)^2 \times (9)^2 \times (2)^4}{4 \times (0.25)^2 \times (5.5)^2} \times 10^{-7} = 1.6.$$

The area of the aperture of the counter is 4 $cm^2 = 4 \times 10^{-4}$ m^2. Therefore the number of particles entering the counter

$$= 1.6 \times (4 \times 10^{-4}) = 6.4 \times 10^{-4}/\text{sec}.$$

4. *In Rutherford scattering experiment the number of particles observed at an angle of 10° is one million/rain, the number of particles per minute will be observed at 90° and at 180° is explained below.*

Solution : By Rutherford formula, the number of particles of a given energy scattered at angle ϕ from a given foil is

$$N(\phi) \propto \frac{1}{\sin^4(\phi/2)} = K \frac{1}{\sin^4(\phi/2)}$$

where K is constant. Now $N(10°) = 10^6$/min.

Thus $$10^5 = K \frac{1}{\sin^4\left(\frac{10}{2}\right)^\circ} = \frac{K}{(0.0872)^4}$$

so that $$K = (0.0872)^4 \times 10^6.$$

Hence the particles observed at 90° are

$$N(90°) = K \frac{1}{\sin^4\left(\frac{90}{2}\right)^\circ}$$

$$= (0.0872)^4 \times 10^6 \times \frac{1}{\left[\left(1/\sqrt{2}\right)\right]^4}$$

$$= 232.$$

Similarly, $N(180°) = 58.$

5. *The distance of closest approach of α-particles to the copper nucleus, when α-particles of 5 MeV energy are scattered back by a thin sheet of copper (Z = 29) is calculated as follows.*

Solution : An α-particle will have its closest approach when its impact parameter is zero, *i.e.*, when it moves directly (head-on) toward the nucleus and is *reflected* back (180° scattering).

Let r_0 be the distance of the α-particle from the nucleus at the closest approach. The (positive) charge on the nucleus is Ze, and that on the α-particle is $2e$, where e is the electronic charge. Therefore, the electrostatic potential energy of the particle at the instant of closest approach is $\frac{1}{4\pi\epsilon_0}\frac{(Ze)(2e)}{r_0}$

At this instant the α-particle is momentarily at rest, and its initial kinetic energy E_k is entirely converted into electrostatic potential energy. Hence at this instant ; we have

$$E_k = \frac{1}{4\pi\epsilon_0}\frac{2Ze^2}{r_0}.$$

$$\therefore \qquad r_0 = \frac{1}{4\pi\epsilon_0}\frac{2Ze^2}{E_k}$$

Here $E_k = 5\text{ MeV} = 5 \times 10^6\text{ eV} = (5 \times 10^6) \times (1.6 \times 10^{-19}\text{ joule})$;

$Z = 29$ and $e = 1.6 \times 10^{-19}$ coul. Also $\frac{1}{4\pi\epsilon_0} = 9 \times 10^9\text{ nt-m}^2/\text{coul}^2$.

$$\therefore \qquad r_0 = 9 \times 10 \times \frac{2 \times 29 \times \left(1.6 \times 10^{-19}\right)^2}{5 \times 10^6 \times 1.6 \times 10^{-19}}$$

$$= 1.67 \times 10^{-14}\text{ meter.}$$

(*b*) The (positive) charge on the proton is e (*i.e.* half of that on α-particle). Therefore the electrostatic potential energy at the instant of closest approach is $\frac{1}{4\pi\epsilon_0}\frac{(Ze)(2e)}{r_0}$. Thus, as above

$$E_k = \frac{1}{4\pi\epsilon_0}\frac{Ze^2}{r_0}$$

$$\therefore \qquad r_0 = \frac{1}{4\pi\epsilon_0}\frac{Ze^2}{E_k}$$

which is half of that of an α-particle of the same energy. Hence for the proton, we have

$$r_0 = \frac{1.67 \times 10^{-14}}{2} = 0.835 \times 10^{-14} \text{ meter.}$$

When an energetic α-particle moves directly (head-on) towards a nucleus, it experiences an increasing electrostatic force of repulsion in accordance to the inverse-square law. Hence its velocity goes on decreasing and on reaching a certain minimum distance from the nucleus the particle comes momentarily to rest. It is then accelerated backwards, reverses its motion and is said to be scattered through 180°. The minimum distance r_0 to which the incident α-particle approaches the nucleus is called the 'distance of closest approach, and is related to the initial kinetic energy E_k of the α-particle by

$$r_0 = \frac{1}{4\pi\epsilon_0}\frac{2Ze^2}{E_k}.$$

The greater the initial kinetic energy E_k of the particle, the smaller the distance of closest approach r_0 . If, however, the energy E_k exceeds a certain maximum value E, the α-particle approaches the nucleus so much closely that the nuclear size becomes comparable with the distance of the particle from it and the inverse-square law breaks down. Besides this, some short-range *attractive* nuclear forces also begin to operate. Hence under this condition the α-particle is not scattered back. The distance of closest approach of a particle of maximum kinetic energy E gives the upper limit of the radius of the nucleus. Thus if the maximum nuclear radius be R, we have

$$R = \frac{1}{4\pi\epsilon_0}\frac{2Ze^2}{E}$$

or

$$E = \frac{1}{4\pi\epsilon_0}\frac{2Ze^2}{R}.$$

Here $Z = 79 \; ; \; e = 1.6 \times 10^{-19}$ coul

and $R = 0.5 \text{ Å} = 0.5 \times 10^{-10}$ meter.

$$\therefore \quad E = (9 \times 10^{-9}) \times \frac{2 \times 79 \times \left(1.6 \times 10^{-19}\right)^2}{0.5 \times 10^{-10}}$$

$$= 7.3 \times 10^{-16} \text{ joule.}$$

Speed of Fastest aα-particle suffering 180°-Scattering: Let E_k be the kinetic energy of the fastest α-particles. The distance of the closest approach of the most energetic (fastest) a-particles suffering 180°-scattering from a nucleus is the maximum possible radius of that nucleus. If the maximum radius of the nucleus (charge + Ze) is R, then the electrostatic potential energy of the α-particle (charge $+2e$) at the instant of closest approach will be $\frac{1}{4\pi\epsilon_0} \frac{2Ze^2}{R}$. Since at this instant the entire kinetic energy E_k has been converted into the potential energy; we have

$$E_k = \frac{1}{4\pi_0}, \frac{2Ze^2}{R}$$

But $E_k = \frac{1}{2} m_\alpha v^2$, where m_α is the mass and v the speed of the fastest particle. Also, given that $m_\alpha = 4m_y$, where m_p is the mass of a proton. Thus

$$E_k = \frac{1}{2} (4m_p) v^2 = \frac{1}{4\pi\epsilon_0} \frac{2Ze^2}{R}$$

or
$$v^2 = \frac{1}{4\pi\epsilon_0} \frac{Ze^2}{m_p R}$$

$$\therefore \quad v = \sqrt{\left(\frac{Ze^2}{\left(4\pi\epsilon_0\right) m_p R}\right)}$$

An α-particle is scattered back (180°) at its closest approach to the nucleus. The distance of closest approach is the upper limit of the nuclear radius. If it is R, then we have

$$R = \frac{1}{4\pi\epsilon_0}\frac{2Ze^2}{E_k}$$

$$= \left(9 \times 10^9 \frac{N-m^2}{C^2}\right)\frac{2\,(79)\left(1.6 \times 10^{-19}\ C\right)^2}{4 \times 1.6 \times 10^{-13}\ J}$$

$$= 5.7 \times 10^{-14}\ \text{m}.$$

$\therefore$ maximum volume in which the positive charge may be concentrated

$$= \frac{4}{3}\pi R^3$$

$$= \frac{4}{3}\ (3.14)\ (5.7 \times 10^{-14}\ \text{m})^3 = 7.75 \times 10^{-40}\ \text{m}^3.$$

6. A beam of protons and α-particles, each of energy 15 MeV is incident on a thin foil of tin (Z = 50). It is observed that the α-particles are scattered backward but not the protons. This observation explained and the maximum possible value of the nuclear radius is found. As to why electrons are not effective in scattering α-particles and protons are also explained.

Solution : The distance of closest approach of α-particle (charge + 2*e*) to a nucleus (charge +Ze) is given by

$$r_0 = \frac{1}{4\pi\epsilon_0}\frac{2Ze^2}{E_k},$$

where E_k is the initial kinetic energy of the incident α-particle. The particle on reaching at this distance from the nucleus is reflected back (scattered 180°).

Here E_k= 15 MeV = 15x 10^6 eV = 15x $10^6 \times 1.6 \times 10^{-19}$ joule.

$$\therefore \qquad r_0 = (9 \times 10^9) \times \frac{2 \times 50 \times \left(1.6 \times 10^{19}\right)^2}{15 \times 10^6 \times 1.6 \times 10^{-19}}$$

$$= 9.6 \times 10^{-15}\ \text{meter}.$$

If 15 MeV is the maximum energy of the α-particles for which they are scattered backwards, then 9.6×10^{-15} meter is also the maximum possible value of the radius of the tin nucleus.

The charge on a proton is + *e*, *i.e.* half of that on an α-particle. Hence a proton of 15 MeV energy will have its distance of closest approach half of that of α-particle *i.e.* 4.8×10^{-15} meter. This means that the proton of this energy will penetrate the nucleus, and not be scattered backward.

The electrons, because of their little mass, are unable to deviate the α-particles and protons appreciably.

7. *With gold foil (Z=79), 180-scattering is observed with α-particles of velocity upto 16 × 10⁷ m/sec. The upper limit for radius of the nucleus is deduced. The mass of an α-particle is 4m*$_p$ *where m*$_p$ *is the mass of proton (1.67 ×10⁻²⁷ kg).*

Solution : The distance of closest approach of an α-particle suffering 180° scattering from a nucleus of atomic number Z; given by

$$r_0 = \frac{1}{4\pi\epsilon_0}\frac{2Ze^2}{E_k}$$

where E_k is the initial kinetic energy of the particle. If m_α be the mass and v the velocity of the particle, then $E_k = \frac{1}{2} m_\alpha v^2$, so that

$$r_0 = \frac{1}{4\pi\epsilon_0}\frac{4Ze^2}{m_\alpha v^2}.$$

If v be the velocity of the *fastest* particle that can suffer 180° scattering, then r_0 is also the upper limit for the radius of the nucleus.

Putting the given values; the upper limit of gold nucleus is

$$r_0 = \frac{\left(9 \times 10^9\right) \times 4 \times 79 \times \left(1.6 \times 10^{-19}\right)^2}{\left(4 \times 1.67 \times 10^{-27}\right) \times \left(1.6 \times 10^7\right)^2}$$

$$\approx 4.26 \times 10^{-14} \text{ meter.}$$

6

The Spectrum

Series Relationship

Spectrum: When a beam of white light is passed through a prism (or grating) it breaks up into the beams of constituent colours. The different coloured beams, when focussed on a screen by a converging lens, form an array of colours on the screen, This is called the 'spectrum' of light.

There are two main types of spectra : (*i*) 'emission spectra' and (*ii*) 'absorption spectra'.

When light coming directly from a source is allowed to enter a prism or grating spectroscope, an 'emission spectrum' of the source is observed in the spectroscope. On the other hand, when light from a source showing a continuous emission spectrum is passed through an absorbing material and then into the spectroscope, an' absorption spectrum of the material is observed.

Emission and absorption spectra are further classified according to their appearance. There are three classes :

Continuous Spectra : When the emitting source is an incandescent solid, or liquid, such as a lamp filament or a gas at high pressure, the spectrum is continuous containing all colours (wavelengths) from red to violet. Its appearance is like an unbroken luminous band of light in which the colour changes gradually from point to point but without any sharp boundary. In this spectrum the intensity is maximum at a certain point and decreases on both sides of it. The point of maximum intensity shifts towards the violet end of the spectrum as the temperature of the source increases.

Line Spectra : If the light source is a low-pressure gas (as in a discharge tube), flame, arc or spark, the spectrum is discontinuous, showing a number of sharp bright-coloured lines. These lines are the images of the slit of the spectroscope formed by lights of different colours. The entire series of images is called a 'line spectrum.' The different lines differ in intensity and nature. Some are sharp, some are sharp on one side and diffused on the other, and others are diffused on both sides.

The line spectrum is the characteristic of the atom or the ion. It means that a particular atom or ion always gives the characteristic set of spectral lines, and no two atoms or ions can give the Atomic and Nuclear same spectral line. For example, sodium atom gives two intense yellow lines called D_1 and D_2 lines.

Band Spectra : Such spectra are obtained by the radiation from gas molecules such as oxygen (O_2), nitrogen (N_2), cyanogen (CN), etc. They consist of illuminated regions, called 'bands', separated by dark spaces. With a high-resolving instrument each band is seen to consist of very fine lines which become closer and closer on one side of the band until they coincide. This side thus has a sharp and bright edge called the 'head' of the band.

Absorption Spectra : The absorption spectra are obtained when the absorbing substance is placed between a source emitting a continuous spectrum and the slit of the spectroscope. In such cases certain colours (wavelengths) are absorbed by the substance. Hence the spectrum is found to consist of dark lines or bands against a bright background. An example of such spectra is sun's spectrum. It is a line absorption spectrum. It was studied in detail by Fraunhoffer who named the absorption lines as *A,B,C,D......* These lines inform us about the elements which are present around the sun.

Similarly, when an intense beam of continuous white light is passed through sodium vapour and then sent into a spectroscope, we obtain two dark lines on a continuous background in the same positions as the yellow D_1 and D_2 lines in the sodium emission spectrum. If the sodium vapour is replaced by Iodine vapour (l_2), an absorption band spectrum of I_2 molecule is obtained.

Series Relationship in Atomic Spectra—The atomic spectra consist of a large number of lines. A quantitative experimental study of these spectral lines was made in the second half of the 19th century when several regularities were observed in the pacings of the lines. For example, in 1870, Liveing and Dewar noticed that the spectral lines of various elements could be grouped into distinct 'series'. *In each series the spacing and intensity of lines decrease regularly towards shorter wavelengths,* until it becomes impossible to distinguish the individual lines. The point at which the lines of the series finally converge is called the 'series limit'. The various series in complicated spectra overlap.

The simplest atomic spectrum is that of hydrogen. Its visible part consists of a single series which was first observed by Balmer in 1885. This is called the 'Balmer series' of hydrogen. Its first line having longest wavelength of 6563 A is named H *a*, the next Hβ , and so on, the series limit reaching at 3646 A. Besides this, there is a series of lines in the ultraviolet part of the hydrogen spectrum which is known as 'Lyman series;

and three series in the infrared part which are known as 'Paschen series', 'Brackett series' and 'Pfund series'.

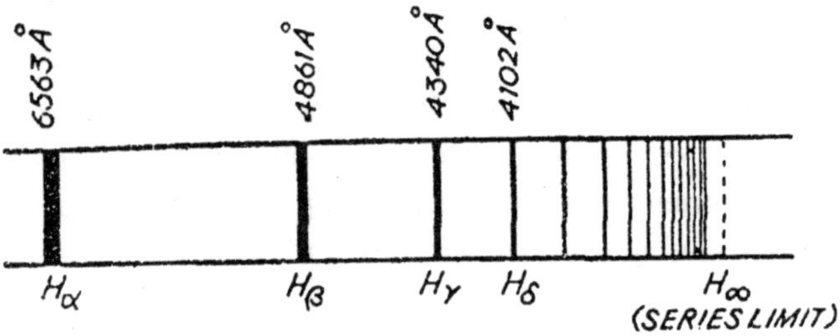

(BALMER SERIES OF HYDROGEN)

Balmer discovered a formula for the wavelengths of all the lines of the Balmer series. His formula is

$$\lambda = 3646 \frac{n^2}{n^2 - 4} \text{ Å} \; ; n = 3, 4, 5, \ldots \ldots$$

$n = 3$ gives the wavelength of H_α line, $n = 4$ gives H_β line, ... and $n = \infty$ gives the series limit.

Subsequently, Rydberg found that Balmer's formula was a special case of a more general formula which is as follows:

$$\bar{\nu} = \frac{1}{\lambda} = R_{\mathrm{H}}\left(\frac{1}{m^2} - \frac{1}{n^2}\right),$$

where $\bar{\nu}$ is wave number (reciprocal of wavelength), m and n are positive integers ($n > m$) and R_H is the Rydberg constant for hydrogen. Its value is 1.097×10^7 meter^{1}. To obtain formula for Lyman series we set $m = 1$ and $n = 2, 3, 4, \ldots$; for Balmer series $m = 2$ and n —3, 4, 5,......; and so on. Thus

$$\bar{\upsilon} = R_H \left(\frac{1}{1^2} - \frac{1}{n^2}\right) \; ; n = 2, 3, 4, \ldots \ldots \qquad \text{(Lyman)}$$

$$\bar{\upsilon} = R_H \left(\frac{1}{2^2} - \frac{1}{n^2}\right) \; ; n = 3, 4, 5, \ldots \ldots \qquad \text{(Balmer)}$$

$$\bar{\upsilon} = R_H \left(\frac{1}{3^2} - \frac{1}{n^2}\right) \; ; n = 4, 5, 6, \ldots \ldots \qquad \text{(Paschen)}$$

$$\overline{v} = R_H \left(\frac{1}{4^2} - \frac{1}{n^2}\right) \; ; n = 5, 6, 7, \ldots \ldots \qquad \text{(Brackett)}$$

$$\overline{v} = R_H \left(\frac{1}{5^2} - \frac{1}{n^2}\right) \; ; n = 6, 7, 8, \ldots \ldots \qquad \text{(Pfund)}$$

We see that the wave numbers of hydrogen lines can be expressed as differences of two terms of the form $\frac{R_H}{n^2}$.

The next simplest spectra are of the 'monovalent' atoms of alkali metals Li, Na, K, etc. The lines in the spectrum of an alkali atom can be grouped into four distinct series ; a 'principal series' of intense lines, a 'sharp series' of fine lines, a 'diffuse series' of comparatively broader lines and a 'fundamental series' which lies in the infrared region. The sharp and diffuse series lie in the visible part and converge to a common limit. Rydberg represented the lines of a particular series by the formula

$$\overline{v} = Z^2 R_A \left[\frac{1}{(m - \Delta_1)^2} - \frac{1}{(n - \Delta_2)^2}\right].$$

where R_A is the Rydberg constant for a particular element A, Z is the atomic number, m and n are positive integers, and Δ_1 and Δ_2 are constants for the particular series. Actually each line of an alkali spectrum is a close doublet. Again we find that the wave numbers of the lines can be expressed as differences of two terms like $\frac{Z^2 R_A}{(m - \Delta)^2}$

After alkali spectra, next in complexity are the spectra of 'divalent' atoms of alkalineearths Be, Mg, Ca. In a typical alkaline-earth spectrum we can distinguish two distinct systems of lines : a system of singlets and a system of triplets. Each system has a principal series, a sharp series, a diffuse series and a fundamental series.

As we proceed to atoms having several valence electrons, the spectra become more complex and the groupings of lines into series become less pronounced. Still regularities can be observed in complex spectra, and it is possible to express the wave number of any spectral line as the difference of two terms.

Rydberg-Ritz Combination Principle

This principle states that *"the wave numbers of spectral lines can be expressed by the difference of spectroscopic terms in such a way that other differences of those terms give also the wave numbers of lines in the same spectrum".*

Suppose in a spectrum the wave numbers of two lines are given as

$$\overline{v}_a = T_2 - T_3 \text{ and } \overline{v}_d = T_1 - T_4,$$

then lines of the following wave numbers are also expected in the same spectrum;

$$\overline{v}_b = T_2 - T_4 \text{ and } \overline{v}_0 = T_1 - T_3 .$$

This means that constant differences exist between the wave numbers:

$$\overline{v}_b - \overline{v}_a = \overline{v}_d - \overline{v}_c$$

$$\overline{v}_c - \overline{v}_a = \overline{v}_d - \overline{v}_b$$

Displacement Law: According to this law 'the spectrum of any neutral atom of atomic number Z closely resembles the spectrum of the singly ionised atom of atomic number Z + 1. For example, the spectrum of He + (Z = 2) closely resembles the spectrum of H (Z = l) and can be represented by a similar formula,

$$\overline{v} = 4 R_{He} \left[\frac{1}{m^2} - \frac{1}{n^2}\right].$$

Other elements deprived of all but one electron, also produce hydrogen-like spectra which can be represented by the general formula

$$\bar{v} = Z^2 R_A \left[\frac{1}{m^2} - \frac{1}{n^2}\right].$$

In a similar way, the spectrum of a singly-ionised alkaline-earth atom resembles the spectrum of an alkali atom. From this we conclude that it is the number of valence electrons in an atom which determines the qualitative character of the spectrum of that atom.

PROBLEMS

1. The wavelength of a yellow line of sodium is 5896 Å. Its wave number and frequency are calculated as follows.

Solution : The wave number $\bar{v}$ is the reciprocal of wavelength λ, that is,

$$\bar{v} = \frac{1}{\lambda}.$$

Here $\lambda = 5896 \text{ Å} = 5896 \times 10^{-8}$ cm.

$$\therefore \quad \bar{v} = \frac{1}{5896 \times 10^{-8} \text{ cm}}$$

$$= 16960 \text{ per cm.}$$

The frequency v is related to the wavelength A by

$$c = v\lambda,$$

where c is the speed of light. Thus

$$v = \frac{c}{\lambda} = \frac{3 \times 10^8 \text{ cm/sec}}{5896 \times 10^{-8} \text{ cm}}$$

$$= 5.1 \times 10^{12} \text{ sec}^{-1}$$

7

The Crystalline

Amorphous Solids

A solid consists of atoms, or ions, or molecules packed closely together. In most of the solids the atoms, ions, or molecules are arranged into regular, orderly and periodic three-dimensional patterns. Such solids are called 'crystals'. Thus a crystal is characterised by having a *long-range order* in its structure. This gives a symmetry to the external shape of the crystals.

Solids which lack regularity in the arrangement of their constituent particles are known as 'amorphous solids'. Such solids may, however, show *short-range order* in their structures. Glass, concrete, paper, pitch and plastic are amorphous solids.

Classification of Crystals

The constituent particles (atoms, ions or molecules) in a crystal are bound together by 'electrostatic' forces, but the distribution of charges in different crystals are qualitatively different. Therefore, the nature of binding varies from crystal to crystal. The difference in bindings is closely connected with the difference in the mechanical, electric and magnetic properties of the crystals.

The binding is characterised by the 'cohesive energy' of the crystal. It is the binding energy per atom or molecule, that is, the energy per atom or molecule that would be liberated in the formation of the crystal from individual neutral atoms or molecules.

The crystals have been classified into five categories according to the binding in them :

1. Ionic Crystals bound by strong electrostatic attraction between ions of opposite kinds.

   Examples: NaCl, CsCl, KBr, etc.

2. Covalent Crystals bound by sharing of electrons between atoms.

   Examples : C (diamond), Ge, Si, etc.

3. Metallic Crystals bound by electrostatic attraction between the lattice of ion cores and the free electron gas.

   Examples : Na, Al, Cu, etc.

4. Inert-gas (or molecular) Crystals bound by weak Van der Waals forces.

   Examples : Solid A, solid CH_4, etc.

5. Hydrogen-bonded Crystals bound by hydrogen bonds.

   Examples : H_2O (ice), HF, etc.

Ionic Crystals

In ionic crystals, electrons are transferred from one type of atoms (which lose electrons readily) to the other type (which have high affinity for electrons). Thus the atoms become positive and negative ions. These ions arrange themselves in a configuration in which Coulomb attraction between ions of opposite signs is stronger than the Coulomb repulsion between ions of the same sign.

Thus the "ionic bond" results from the electrostatic interaction of oppositely charged ions. The configuration remains stable in spite of the net electrostatic attraction between the ions because electron shells of different ions cannot mesh together due to the forces present by the action of the exclusion principle.

Examples : The examples of ionic crystals are sodium chloride (NaCl), cesium chloride (CsCl), lithium fluoride (LiF), potassium bromide (KBr), etc.

In the NaCl crystal the outermost (3*s*) electron of the sodium atom ($1s^2 2s^2 2p^6 3s$) is transferred to the 3*p* subshell of the chlorine atom ($1s^2 2s^2 2p^6 3s^2 3p^5$). The Na^+ and Cl~ ions so formed are arranged in a pattern shown in Fig. below. The ions of either kind are located at the corners and at the centres of the faces of an assembly of cubes. Such a crystal structure is called 'face-centered cubic'.

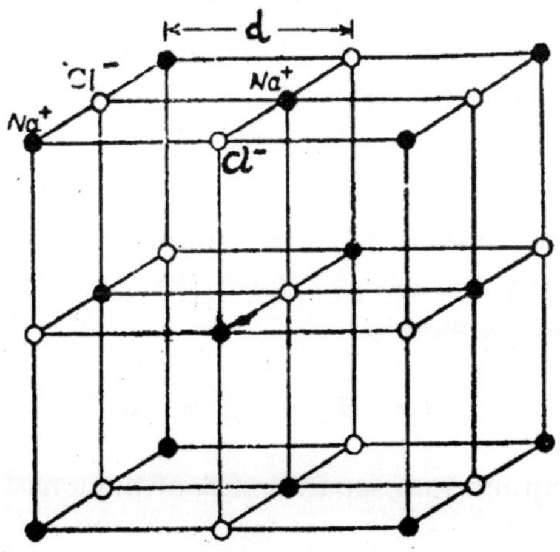

Cohesive Energy

The cohesive energy of an ionic crystal is the energy that would be liberated in the formation of the crystal from individual neutral atoms, or the energy needed to break the crystal into individual atoms. Let us calculate it:

The Coulomb energy of attraction between two ions separated by a distance r is

$$U_{att} = -\frac{e^2}{4\pi\epsilon_0 r}.$$

In the crystal each ion interacts with all its neighbouring ions of same as well as of opposite signs. To take this into account, the net energy of attraction per ion-pair may be written as

$$U_{att} = -\alpha \frac{e^2}{4\pi\epsilon_0 r},$$

where α is called the 'Madelung constant' of the crystal. For crystal structures, α lies between 1.6 and 1.8.

As mentioned above, two ions cannot continuously approach each other under Coulomb attraction on account of the exclusion principle. When they are at a certain small distance apart, they begin to repel each other with a force which rises very rapidly with the decrease in distance. The energy of this (non-electrostatic) repulsion can be expressed as

$$U_{rep} = \frac{B}{r^n}$$

where B is a constant and n is a number ($n \approx 9$). The total potential energy per ion-pair due to their interactions with all the other ions is therefore

$$U = U_{att} + U_{rep} = -\alpha\frac{e^2}{4\pi\epsilon_0 r} + \frac{B}{r^n}.$$

At the equilibrium separation r_0 of the ions U must be a minimum.

This means that at $r = r_0$, $dU/dr = 0$.

Hence

$$\left(\frac{dU}{dr}\right)_{r=r_0} = \frac{\alpha e^2}{4\pi\epsilon_0 r_0^2} - \frac{nB}{r_0^{n+1}} = 0$$

or
$$B = \frac{\alpha e^2}{4\pi\epsilon_0 n} r_0^{n-1}.$$

The potential energy at $r = r_0$ is thus

$$U = -\alpha\frac{e^2}{4\pi\epsilon_0 r_0} + \frac{\alpha e^2}{4\pi\epsilon_0 n r_0}$$

or
$$U = -\frac{\alpha e^2}{4\pi\epsilon_0 r_0}\left(1 - \frac{1}{n}\right).$$

This is the total potential energy per ion-pair in the crystal. In NaCI crystal, the equilibrium distance between ions is $r_0 = 2.81$Å, and $\alpha = 1.748$. Taking $n = 9$, the potential energy per ion-pair is

$$U = -\frac{\alpha e^2}{4\pi\epsilon_0 r_0}\left(1 - \frac{1}{n}\right)$$

$$= -\frac{\left(9\times10^9 \text{ newton-m}^2/\text{coul}^2\right)\times 1.748\times\left(1.6\times10^{-19}\text{coul}^2\right)}{\left(2.81\times10^{-10}\text{ m}\right)}\left(1-\frac{1}{9}\right)$$

$$= 1.27 \times 10^{-18} \text{ joule}$$

$$= -7.97 \text{ eV}.$$

Hence the contribution *per ion* to the cohesive energy of the crystal is half of this, *i.e.*, – 3.99 eV. This is called the 'lattice energy per ion' in the NaCI crystal.

In fact, some energy is needed to transfer an electron from a Na atom to a CI atom to form $Na^+ - CI^-$ ion pair. This is equal to the difference between the + 5.14 eV ionisation energy of Na and the – 3.61 eV electron affinity of Cl, *i.e.*, +1.53 eV. Each atom thus contributes + 0.77 eV to the cohesive energy. The total cohesive energy per atom in the NaCl crystal is therefore

$$U_{cotesive} = -3.99 + 0.77 = -3.22 \text{ eV/atom.}$$

This is greater than the binding energy per atom (2.1 eV/atom) in an actual NaCl molecule.

Properties : An ionic bond is a strong bond because the binding forces are of electrostatic origin. In view of the cohesive energies of ionic solids, which are of the order of several electron-volts, we expect the following properties of these solids:

(i) Most ionic solids are hard cubic crystals, owing to the strength of the bonds between their constituent ions.

(ii) They are usually brittle, since the slipping of atoms past one another (which accounts for the ductility of metals) is prevented by the ordering of positive and negative ions.

(iii) They have high melting and vaporizing points because of their cohesive energies of the order of several electron volts. In order to vaporise an ionic solid we must heat it to a temperature T such that $kT \approx 1$ eV. Obviously, T must be quite high.

(iv) They are transparent to visible radiation. A photon of visible light will interact with a material only which it can excite an electron from its ground state to an excited state. Since the ionic solids all have filled shells, an electron must be excited *from one shell to the next.* The energy required for this is much larger the visible-photon energy. So the visible photons pass right through ionic solids.

(v) They absorb strongly in the infrared. This is because an ion in a solid is capable of oscillation with infrared frequencies"

(vi) They are poor conductors of electricity, because they do not have free electrons. However, in contrast to metals, the conductivity of ionic solids increases with rise in temperature.

(vii) Ionic solids are usually soluble in polar liquids like water. The water molecule has an electric dipole moment, which exerts an attractive force on the charged ions, breaking the ionic bonds and dissolving the solid.

(viii) Ionic solids do not distinguish one direction in space from another. This is because the *filled* shells of the ions are spherically symmetric and so the ionic bonds are *not* directional.

Covalent Crystals

In covalent crystals, one or more electrons are detached from two adjacent atoms and are *shared* equally by both of them. The simplest example of electron sharing is the H_2 molecule in which two electrons are shared by the two atoms (Fig.). As these electrons circulate, they spend more time between the atoms (in fact between the protons) than elsewhere and this produces an attractive force.

The two electrons are said to form a 'covalent' bond. The spin of the electrons in the bond are antiparallel, *i.e.*, the electrons form a pair.

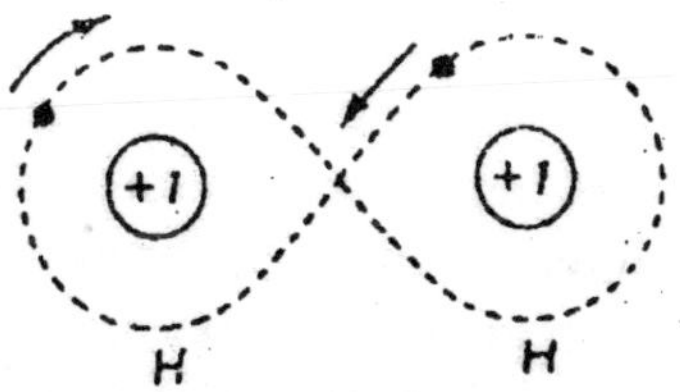

Examples : Diamond (C), silicon (*Si*), germanium (*Ge*), gray tin (*Sn*), silicon carbide (*SiC*) and zinc sulphide (ZnS) are the examples of covalent crystals.

The below given figure shows how the solid structure of diamond is constructed. A central carbon atom is bound to four close neighbours at tetrahedral angles, with each of which it shares electrons in covalent bonds. This happens with each carbon atom in the extended structure. Silicon and germanium also have a diamond like structure. In SiC each atom is surrounded by four atoms of the other kind in the same tetrahedral structure as diamond. (Fig.).

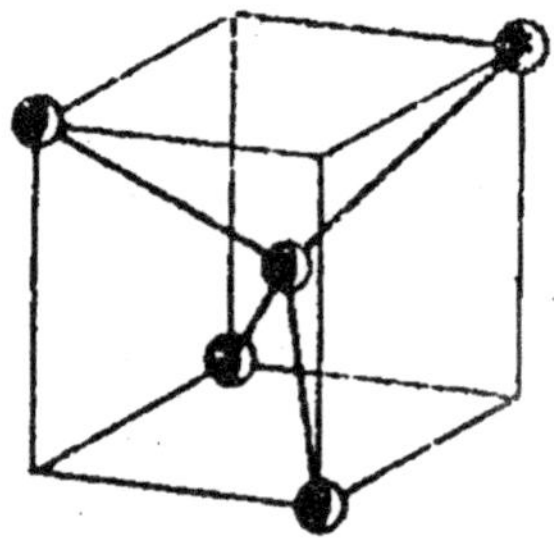

Properties : A covalent bond is a strong bond. Cohesive energies of 6 to 12 eV/atom are typical of covalent crystals. As an example, the bond between two carbon atoms in diamond has a cohesive energy of 7.3 eV with respect to separated neutral atoms. This is comparable with the cohesive energies (bond strengths) in many ionic crystals. As such there is no sharp distinction between ionic and covalent crystals. In fact there is a continuous range of crystals between the ionic and covalent limits.

The actual properties of a given covalent crystal depends on its bond energy. In general, covalent crystals with large bond energies are very hard, have high melting points, and are transparent to visible light. Those with small bond energies may have very different properties. For example, germanium and tin are soft (compared to diamond), melt at much lower temperatures and are reflective to visible light.

The conductivity of covalent crystals varies over a wide range, and increases with increasing temperature.

In contrast to ionic crystals, the covalent crystals have strong *directional* properties.

Metallic Crystals

Metallic crystals, commonly known as metals, are made up of atoms whose valence electrons are weakly coupled to the nucleus: In the solid state, the atoms are so close that they give up their valence electrons which wander throughout the interior of the metal. Thus *a metal consists of Positive ion cores immersed in a gas of free electrons.* The (metallic) binding results from the attraction between the positive metal ions and the electron gas, which exceeds the mutual repulsion of the electrons in the gas and of the positive ions.

Examples : Most of the elements in the first four groups of the periodic table and the transition elements are metals. Li, Na, Cu, Ag, Zn, Fe are some examples.

Some elements are on the borderline between metallio and covalent crystals. Tin is an example. Above 13.2° C the metal "white tin" exists, while below 13.2° C the covalent solid "gray tin" exists.

Metallic crystals exist in bcc, fcc and hcp structures.

Properties : The characteristic properties of metals are their high electrical and thermal conductivities, surface luster, ductility and opacity.

(i) The high conductivity follows from the ability of the free electrons to move throughout the metal, (In contrast, the electrons in ionic and covalent crystals are bound to particular atoms or pairs of atoms).

(ii) When light shines on a metal, the free electrons oscillate under the electromagnetic field of the incident light and become sources of light. This gives the metal its surface luster.

(iii) Since the metallic bonds are not localised between adjacent atoms, the atoms of a metal can be re-arranged in position without rupturing the crystal. This *explains* ductility of metals

(iv) Metallic bonds are weaker than ionic and covalent bonds. For example, cohesive energy in sodium is 1.13 eV/atom and that in zinc 1.35 eV/atom. As a result, metals interact strongly with photons of visible light, and so metals are not transparent. In fact, they are highly reflective.

(v) Because of the lower cohesive energies, metals have relatively lower melting points. Transition metals, however, fall in the upper limit of cohesive energy and so they have higher melting points.

(vi) Since metallic bonds do not depend on any sharing or exchange of electrons between specific atoms, the exact nature of the atoms of the metal is not as important as it is in case of ionic or covalent solids. Hence we can make various metallic alloys by mixing different metals in varying proportions.

Molecular or Inert-gas Crystals : Many solids, or crystals, are composed of *neutral* atoms or molecules, without any transferring, sharing or free movement of electrons which could form ionic, covalent or metallic bonds. These solids remain bound by *much weaker, short-range* attractive forces, known as "Van der Waals forces". These forces arise as a result of fluctuations in the charge distributions of nearby molecules.

Inert-gas solids like solid argon and many organic symmetric-molecule solids like solid methane (CH_4) are the examples of molecular solids.

To investigate the origin of Van der Waals forces, let us consider two identical inert-gas atoms at a separation r. The electronic charge distribution in each atom is spherically symmetric *on the average**. However, the electrons are in constant motion around the nucleus. Therefore, *at any instant* one part or other of the atom may have an excess of them, that is, the atom may have an *instantaneous* electric dipole moment. As shown in below given figure for two instants, an instantaneous dipole moment of magnitude p_1 on one atom produces an electric field E of magnitude $2p_1/r^3$ at the centre of the second atom. This field will induce an instantaneous dipole moment $p_2 = \alpha E = 2\alpha p_1/r^3$ on the second atom, where α is the polarizability of the potential energy of the dipole moments is

$$U(r) = -\frac{2p_1p_2}{r^3} = -\frac{4\alpha p_1^2}{r^6}$$

or
$$U(r) = -\frac{C}{r^6}$$

where C is a constant. The mutual energy of the two atoms arising from their (Van der Waals) interaction is thus *negative,* signifying that the force between them is *attractive.* The force itself is equal to —dU/dr and so is proportional to r^{-7}, which means that it drops rapidly with increasing separation. It is this force which bonds the crystals of inert gases and also the crystals of many organic molecules. Ordinary gases like O_2, N_2, H_2 form molecular solids in the solid state.

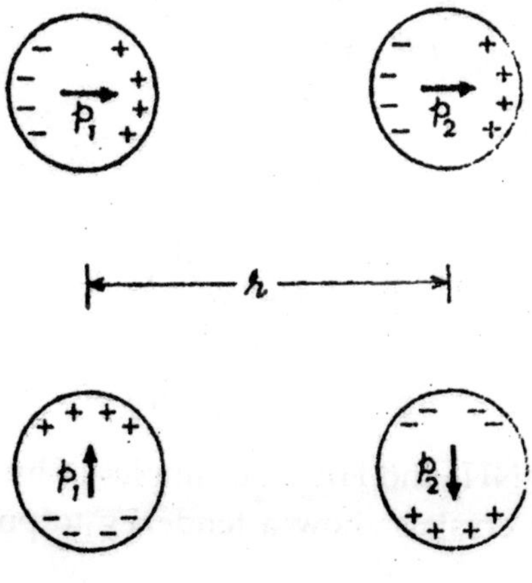

Properties : Van der Waals forces are much weaker than those found in ionic and covalent bonds. This means that little energy is required to break the bonds. Hence the molecular crystals have low melting and boiling points and little mechanical strength. It is easy to deform and compress them. The absence of free electrons makes them very poor conductors of heat and electricity.

Cohesive energies are low, only 0.08 eV/atom in solid argon (melting point —189°C) and 0.1 eV/molecule in solid methane (melting point — 183°C).

Hydrogen-bonded Crystals

In certain crystals an atom of hydrogen attracts two extremely electronegative atoms, particularly F, O and N, thus forming a "hydrogen bond" between them. In the extreme case, the hydrogen atom loses its electron to another atom in the molecular the bare proton forms the hydrogen bond. These bonds are stronger than Van der Waals bonds, but weaker than ionic and covalent bonds.

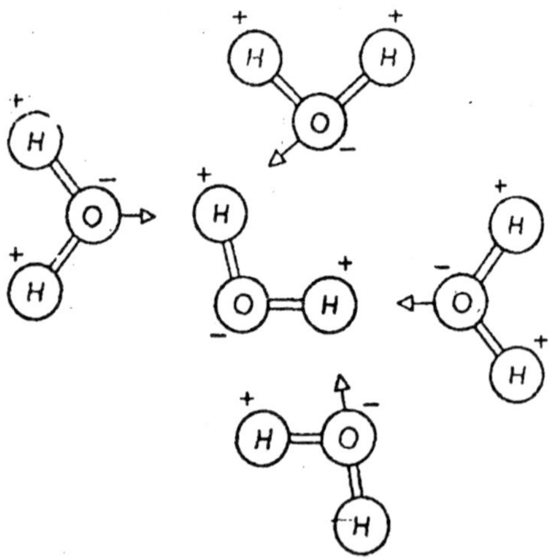

H_2O (ice), NH_3 and HF are examples of hydrogen-bonded crystals. Such crystals show a tendency to polymerise.

In H_2O molecule, the oxygen atom attracts all the electrons of the molecule and becomes the negative end. The two 'bare' protons form the positive ends, and each attracts the negative oxygen of two adjacent molecules (Fig.). The hexagonal pattern of an ice crystal arises from the tetrahedral arrangement of the four hydrogen bonds each H_2O molecule can participate in.

The following crystals are classified according to their bondings

Na, NaCl, diamond, KBr, Cu, Zn, Si.

Ionic : NaCl, KBr.

Covalent : Diamond, Si.

Metallic ; Na, Cu, Zn.

PROBLEM

1. The ionic cohesive energy and the atomic cohesive energy of CsCl are found out as here. The Madelung constant for CsCl is 1.763 and the separation between adjacent ions is 3.56 Å. The non-electrostatic repulsive energy varies as r^{-n}, where $n = 10.5$. The ionisation energy of Cs is 3.89 eV and the electron affinity of Cl is – 3.61 eV.

Solution: The ionic cohesive energy of CsCl is given by

$$U = -\frac{\alpha e^2}{4\pi\epsilon_0 r_0}\left(1 - \frac{1}{n}\right),$$

where $\frac{1}{4\pi\epsilon_0} = 9 \times 10^9$ N-m^2/C^2.

Substituting the given data for CsCl,

$$U = -\frac{(9 \times 10^9\ N - m^2/C^2)(1.763)(1.6 \times 10^{-19}\ C)^2}{3.56 \times 10^{-10}\ m}\left(1 - \frac{1}{10.5}\right)$$

$$= -10.3 \times 10^{-19} \text{ J}$$

$$= -\frac{10.3 \times 10^{-19} \text{ J}}{1.6 \times 10^{-19} \text{ j/eV}}$$

$$= -6.45 \text{ eV}.$$

This represents the energy needed to remove Cs^+ and Cl^- ions from the crystal. To form Cs^+ from Cs atom requires 3.89 eV, and to form Cl^- from neutral Cl atom releases 3.61 eV. Thus, given Cs^+ and Cl^-, we supply 3.61 eV to form Cl^-, and get back 3.89 eV when we form Cs. The energy needed to form the neutral atoms from the crystal is

$$-6.45 \text{ eV} - 3.61 \text{ eV} + 3.89 \text{ eV} = -6.17 \text{ eV}.$$

This is the atomic cohesive energy.

8

Crystal Structure

Crystal structure is explored through the diffraction of waves having a Wavelength comparable with the interatomic spacing (10^{-10} m) in crystals. Radiation of longer wavelength cannot resolve the details of structure, while radiation of much shorter wavelength is diffracted through inconveniently small angles. Usually, diffractions of X-rays, neutrons and, less often, electrons is employed in the study of crystal structure.

W. L. Bragg presented an explanation of the observed diffracted beams from a crystal. He supposed that the incident waves undergo specular (mirror-like) reflection at the *various* parallel planes of atoms in the crystal, with each plane reflecting only a small fraction of the radiation. *The diffracted beams are found only when the reflections from the various planes of atoms interfere constructively.*

In the below given figure is shown a particular set of atomic planes in a crystal, d being the inter-planar spacing. Suppose an X-ray beam is incident at a glancing angle θ. It is scattered by the atoms like A and B in random directions. Constructive interference takes place only between those scattered waves which are reflected specularly and have a path difference of $n\lambda$, where λ is the X-ray wavelength and n is an integer. The path difference for the waves reflected from adjacent planes is

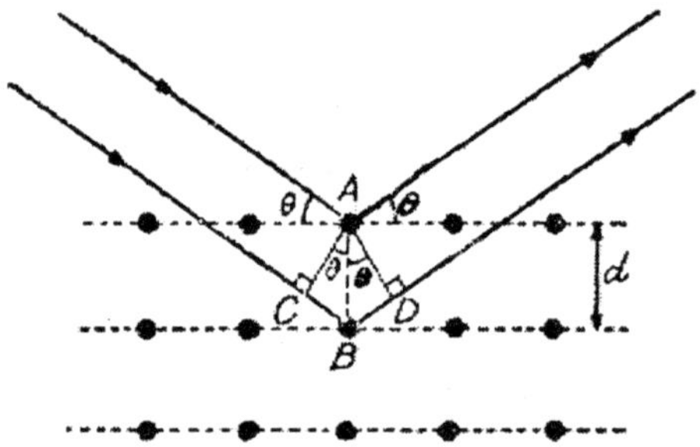

$$CB + BD = d \sin\theta + d \sin\theta = 2d \sin\theta.$$

For constructive interference, we must have

$$\boxed{2d \sin\theta = n\lambda.} \quad n = 1, 2, 3, \ldots\ldots$$

This is Bragg's law. It shows that for given values of λ and d, only for certain values of θ (corresponding to $n = 1, 2, 3, \ldots\ldots$), the reflected waves add up in phase to give a strong reflected (diffracted) beam. The images obtained in these directions are the first, second, third.......order diffraction images.

The Bragg law is a consequence of the periodicity of the space lattice. The arrangement of atoms (composition of the basis) associated with each lattice point determines the relative intensity of the various orders n of diffraction from a given set of parallel planes.

The maximum value of sin θ is 1. Therefore, Bragg reflection can occur only for wavelengths $\lambda \leq 2d$, and d is of the order of 1 Å. This is why we cannot use visible light.

The Bragg law requires that θ and λ should be so related as to satisfy the equation $2d \sin\theta = n\lambda$. X-rays of wavelength A striking a three-dimensional crystal at any angle of incidence will, in general, not be reflected. To satisfy the Bragg law in experiment, either λ or θ is varied.

There are four experimental methods of diffraction used in crystal structure analysis : (*i*) Bragg method, (*ii*) Laue method, (*iii*) rotating crystal method and (*iv*) Powder method.

Investigation of Crystal Structure by Bragg's Method

The Bragg's method of studying the crystal structure consists in finding out the interplanar spacings, d_{hkl}, of the various sets of planes rich in atoms. From a knowledge of d_{hkl}, and from the relative intensities of reflection for different orders, the crystal structure can be deduced.

The experimental arrangement required is a Bragg's X-ray spectrometer (Fig.). The crystal C under investigation is mounted on the spectrometer table so that its (100) face is on the axis of rotation. The position of the table can be read on a circular scale provided with a vernier V_1. A strong, *monochromatic* X-ray beam is first passed through two fine slits S_1 and S_2 and then made to fall on the crystal face. The reflected beam after passing through the silt S_3 enters an ionisation chamber D mounted on an arm which can be rotated about the same axis as the crystal. Its position can be read by a second vernier V_2. The gas in the chamber is ionised by the X-rays. The resulting ionising current, measured by the electrometer E, is a measure of the intensity of X-rays reflected by the crystal.

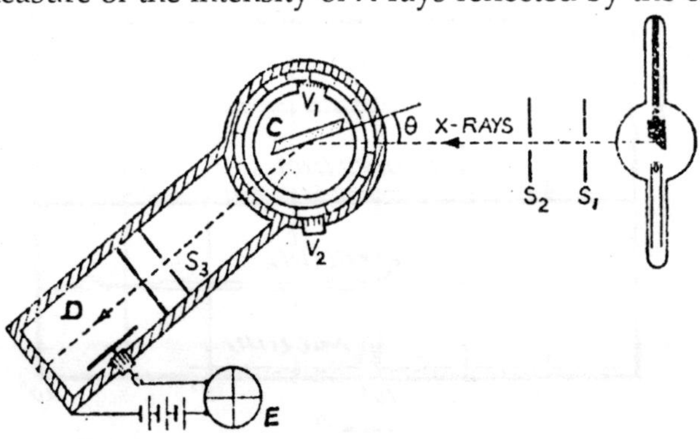

The crystal is rotated through small angles (while the arm carrying the ionisation chamber is rotated through double the angles) and the ionisation current is measured each time. The current is plotted against the glancing angle θ. The curve so obtained shows a number of peaks which correspond to those glancing angles θ which satisfy the Bragg's equation $2d_{100} \sin \theta = n\lambda$ (Fig.). The different peaks correspond to different orders (n = 1, 2, 3,...). Hence the corresponding glancing angles can be directly read from the curve. Now, the (110) face and the (111) face of the crystal are presented turn by turn to the incident beam of X-rays and the same procedure is gone through. Since today the absolute wavelength λ of a given X-radiation is known from measurements with ruled gradings, the interplanar spacings d_{100}, d_{110} and d_{111} can be determined from the Bragg's equation. If, however, the X-ray wavelength is not known, the ratios of the interplanar spacings can be determined.

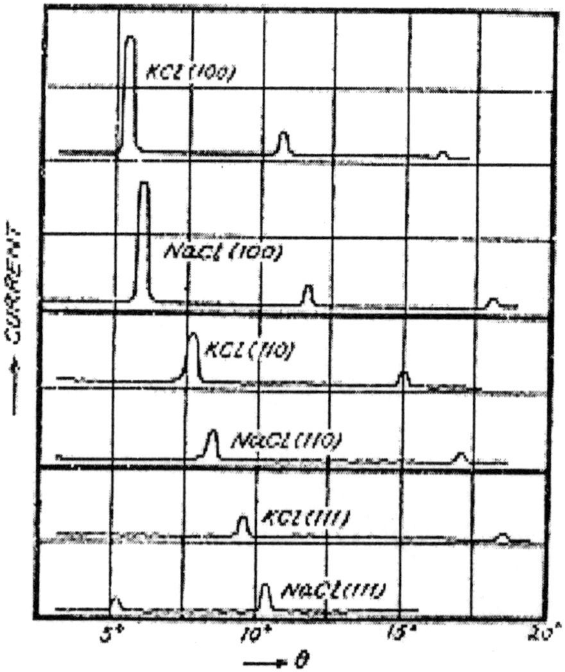

Suppose for a particular crystal used on a Bragg's spectrometer, strong reflections from the sets of planes (100); (110); (111) are obtained for angles $\theta_1, \theta_2, \theta_3$ respectively in the first (n = 1) order. Then, from Bragg's equation, we have

$$2\,d_{100} \sin\theta_1 = \lambda\,,$$

$$2d_{110} \sin\theta_2 = \lambda\,,$$

and

$$2d_{110} \sin\theta_3 = \lambda.$$

Hence $$d_{100} : d_{110} : d_{111} = 1/\sin\theta_1 : 1/\sin\theta_2 : 1/\sin\theta_3.$$

Bragg found that for a KCl crystal ; $\theta_1 = 5°23'$; $\theta_2 = 7°37'$ and $\theta_3 = 9°25'$. Hence for KCl

$$\begin{aligned} d_{100} : d_{110} : d_{111} &= 1/\sin 5°23' : 1/\sin 7°37' : 1/\sin 9°25' \\ &= 1/0.0938 : 1/0.1326 : 1/0.1636 \\ &= 1/1 : 1/1.414 : 1/1.744 \\ &= 1 : 1/\sqrt{2} : 1/\sqrt{3}\,. \end{aligned}$$

Theoretically, this ratio is found to hold for a simple cubic lattice structure. Hence it is concluded that the KCl *crystal has a simple cubic structure*. However, from these data we cannot know the constituents are the lattice corners, whether they are molecules of KCl or atoms (or ions) of K and Cl alternately. This can be decided from a study of other similar crystals such as NaCl.

When the first order reflections from the three faces (100), (110) and (111) of NaCl are compared, the ratios between the interplanar spacings are found as :

$$d_{100} : d_{110} : d_{111} = 1 : 1/\sqrt{2} : 2/\sqrt{3}\,.$$

which agree with the theoretical values for a faced-centered cubic lattice. From this we conclude that the NaCl *crystal has fee lattice* (whereas KCl had shown a simple cubic structure).

There is no reason why should there be a difference in the lattice structures of KCl and NaCI. The observed dissimilarity can be explained if we compare the curves of the two crystals. The curves for the (100) and (110) planes of the two crystals are similar in the sense that the intensity falls off continuously as we go to higher orders of reflection. However, *the curves for the* (111) *planes of the two crystals differ.* In case of NaCI (111) the second order is *stronger* than the first order and similarly, as Bragg found, the fourth order is stronger than the third. Such an alternation of intensity in the reflection from the (111) planes is observed in all alkali halides with the exception of KCl.

Bragg showed that the only way of explaining the intensity alternation is to assume that *the lattice points of the cubic structure are occupied by single ions and not by molecules,* and the ions of the two kinds (Na^+ and Cl^- or K+ and Cl^-) occur alternately, as shown in the figure below. Now it can be seen that the (100) and (110) planes contain equal number of ions of both the kinds. But the (111) planes contain, alternately, all Na^+ (or K^+) ions and all Cl^- ions. That is, the Na^+ (or K^+) planes occur exactly half-way between Cl^- planes. When the glancing angle is correct for the *first order* reflection from the Na^+ (or K^+) planes and also from the Cl^- planes, the rays reflected from a Na^+ (or K^+) plane are exactly out of phase with those reflected from the adjacent Cl^- plane thus causing destructive interference. Therefore, the first order reflection is weakened. For the second order reflection there will be reinforcement between all the reflected rays as seen by substituting in Bragg's equation. In general, for reflection at the (111) face the odd orders will be weaker compared with the even orders *i.e.* there will be an intensity alternation.

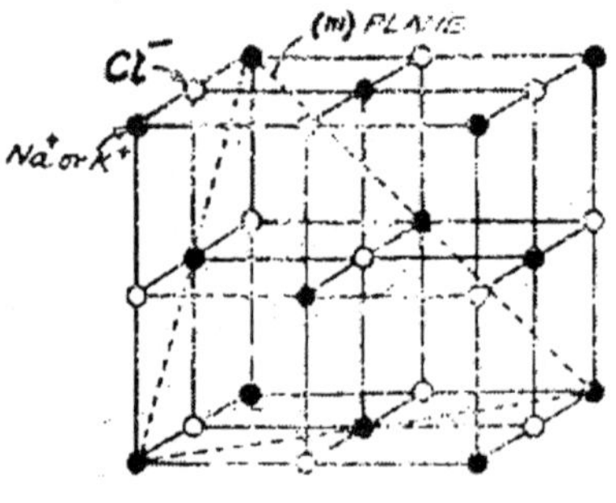

The fact that the intensity alternation is observed in case of NaCl (and other alkali halides) but not in KCl is readily explained when we realise that the reflecting (or scattering) powers of different ionic planes in a crystal are different. In fact, the reflecting power of an atom or ion is proportional to the number of electrons in it *i.e.* to the atomic number Z. Therefore, the Na^+ ions (Z = 10) and Cl^- ions (Z = 18) have quite different reflecting powers so that the rays reflected from the Na^+ and Cl^- planes appreciably differ in intensity and hence the destruction of the odd orders is not complete. That is, *in* NaCl *the odd orders become only weaker* but are still present. On the other hand, the K^+ ions (Z = 18) and Cl^- ions (Z = 18) have equal reflecting powers so that the rays reflected from the K^+ and Cl^- planes have equal intensities and the destruction of the odd orders is complete. That is, in *case of KCl crystal, the odd orders are entirely missing* and the intensity alternation is not observed. Thus in KCl (111) the first observed peak (Fig.), in fact, corresponds to the *second order* reflection (and not to the first order). When this correction is made, we get the same interplanar spacing ratios for KCl as for NaCl *i.e.*

$$d_{100} : d_{110} : d_{111} = 1 : 1/\sqrt{2} : 2/\sqrt{3}.$$

which correspond to an *fcc* lattice. The Na^+ (or K^+) ions alone form an *fcc* lattice, and so do the Cl^- ions. If the reflecting powers of the two kinds of ions are the same, the distinction between them disappears and the structure reduces to a simple cubic lattice as was first found in case of KCl.

Once the arrangement of atoms (or ions) in the, cubic crystal has been decided, we can calculate the absolute value of interplanar spacing.

Laue Method

In Laue's method, a single crystal is held *stationary* in a *continuous* X-ray beam. The crystal diffracts the discrete values of λ for which the crystal planes of spacing d and the incidence angle θ satisfy the Bragg's law.

The experimental arrangement is shown in Fig. (*a*). A continuous (≈ 0.2 Å to 2 Å) *X*-ray beam, well-collimated by a pinhole arrangement, is allowed to fall on the crystal. The crystal holder is adjusted to obtain proper orientation of the crystal. A flat film *A* is placed to receive the transmitted-diffracted beams (or a flat film *B* to receive the reflected-diffracted beams). The diffraction pattern is a regular one, consisting of a series of spots and is characteristic of crystal structure. It is called the Laue pattern (Fig *b*).

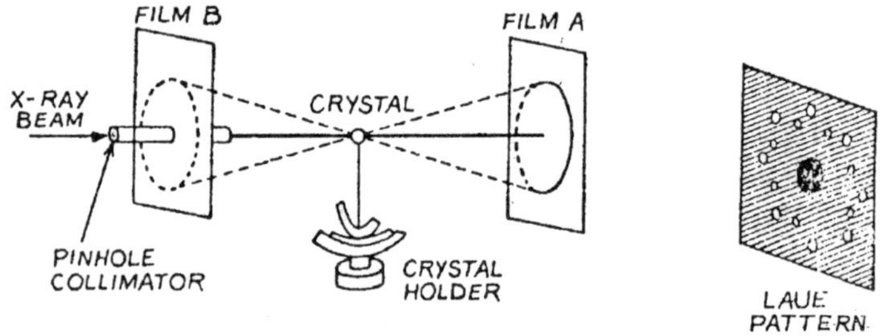

(a) (b)

Each spot in the Laue pattern corresponds to an interference maximum for a set of crystal planes satisfying the Bragg equation $2d \sin \theta = n\lambda$ for a particular wavelength selected from the (continuous) incident beam. The atomic arrangement in the crystal can, therefore, be deduced from a study of the positions and intensities of the Laue spots.

The distribution of spots in the Laue pattern depends on the symmetry of the crystal and its orientation with respect to the *X*-ray beam. For example, if a crystal with four-fold axial symmetry is oriented with the axis parallel to the beam, then, the Laue pattern will show the four-fold symmetry, as in Fig. *b*. The Laue method is convenient for the rapid determination of crystal orientation and symmetry which is needed for solid state experiments. It is also used to study crystalline imperfections under mechanical and thermal treatment. The method is, however, not used for actual crystal structure determination. This is because it is possible for several wavelengths to reflect in different orders from the same plane, so that different orders may superpose on a single spot. This would make the analysis difficult.

Rotating Crystal Method

In this method, a single crystal is rotated about a fixed axis in a *monochromatic* beam of X-rays (or neutrons). The variation in the angle θ presents different atomic planes of the crystal to the X-ray beam for reflection.

A simple experimental arrangement is shown in Fig. below. The crystal under investigation is mounted on a rotating spindle, and the film is wrapped on a cylindrical surface coaxial with the spindle. The incident X-ray beam, made monochromatic by a filter or by reflection from an earlier crystal, is diffracted from a given crystal plane whenever in the course of rotation the value of *6* satisfies the Bragg equation. The beams reflected from all planes parallel to the (vertical) axis of rotation lie in the horizontal plane, while those from planes with other orientations lie above and below the horizontal plane.

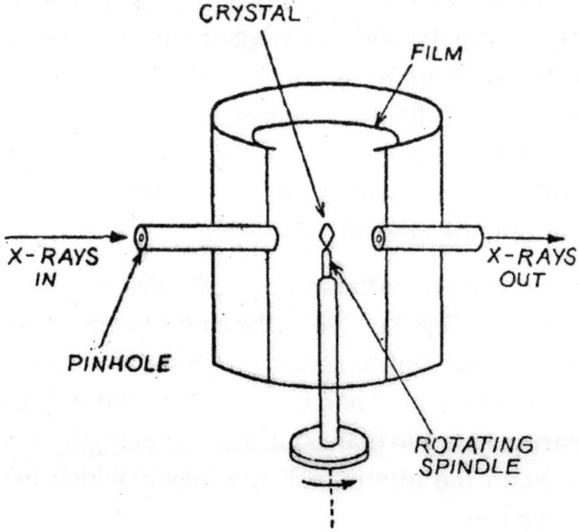

Modern rotating-crystal diffractometers use scintillation counters or proportional counters to detect the diffraction radiation. They automatically collect all the data regarding crystal structure.

Crystal Analysis by Powder Method

It is a more straight forward method for the analysis of crystal structure, developed independently by Debye and Scherrer in Germany and by Hull in America in 1916. In this method, the specimen, instead of being a single crystal, is a finely-powdered crystal.

The minute crystals, called crystallites, of this powder are all randomly (almost continuously) oriented so that they make all possible angles with the incident radiation. Thus all orders of reflection from all possible atomic planes are recorded at the same time.

The experimental arrangement is shown in Fig. below. A fine beam of monochromatic X-rays is allowed to fall on a thin-walled capillary tube *P* containing the powdered crystal. Diffracted rays go out from individual crystallites which happens to be oriented so that a particular family of atomic planes in them make such an angle θ with the incident beam that the Bragg relation $2d \sin \theta = n\lambda$ is satisfied. All such diffracted rays will lie on a conical surface having its apex at *P* and semi-vertical angle 2θ. If a photographic film be kept in the form of a cylinder with its axis perpendicular to the incident beam, one obtains arcs of circles (Fig.).

These arcs are the intersections of cones with the cylinder (film). The central spot is due to the direct beam. Different arcs on one side of the central spot correspond to different atomic planes. From these arcs one can calculate the Bragg angles θ for the various atomic planes. If the wavelength *X* is known, one can obtain the interplanar spacings *d* which lead to the crystal structure.

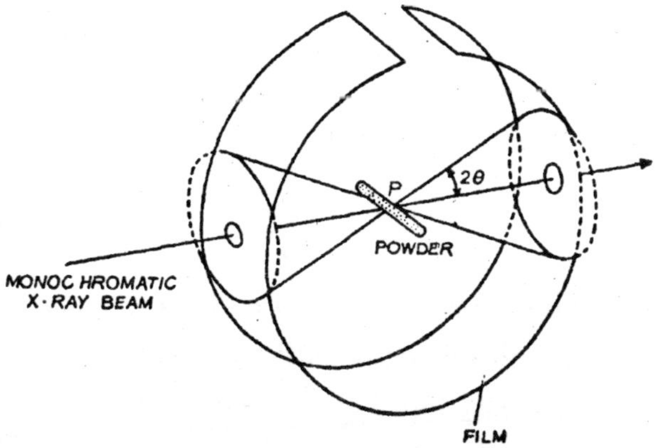

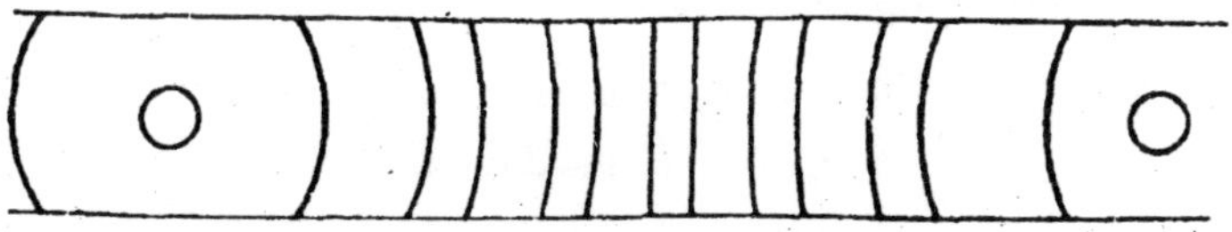

Reciprocal Lattice

In a crystal lattice there exist many sets of planes, with different orientations and spacings, which can cause diffraction. If we draw, from a common origin, normals to all sets of planes, the length of each normal being proportional to the reciprocal of the interplanar spacing of the corresponding set, then the end-points of the normals form a lattice which is called the 'reciprocal lattice'.

Each point in the reciprocal lattice preserves the characteristics of the set of planes which it represents. Its direction with respect to the origin represents the orientation, of the planes, and its distance from the origin represents the interplanar spacing of the planes.

Reciprocal Lattice Vectors : If $\vec{a}, \vec{b}, \vec{c}$ are the translation vectors of a (direct) crystal lattice, then the translation vectors $\vec{A}, \vec{B}, \vec{C}$ of the reciprocal lattice are defined by

$$\left.\begin{aligned}\vec{A} &= 2\pi\frac{\vec{b}\times\vec{c}}{\vec{a}.\vec{b}\times\vec{c}}\\ \vec{B} &= 2\pi\frac{\vec{c}\times\vec{a}}{\vec{a}.\vec{b}\times\vec{c}}\\ \vec{C} &= 2\pi\frac{\vec{a}\times\vec{b}}{\vec{a}.\vec{b}\times\vec{c}}\end{aligned}\right\} \quad \ldots (i)$$

The common denominator* in each case, $\vec{a}.\vec{b}\times\vec{c}$, is the volume of the direct crystal lattice. Thus $\vec{a}, \vec{b}, \vec{c}$ have the dimensions of length, and $\vec{A}, \vec{B}, \vec{C}$ those of $[\text{length}]^{-1}$. From the defining eq. (*i*), $\vec{A}, \vec{B}, \vec{C}$ the axes vectors of the reciprocal lattice, have the following properties :

$$\left.\begin{aligned}&\vec{A}.\vec{a} = 2\pi \quad &&\vec{B}.\vec{a} = 0, \quad &&\vec{C}.\vec{a} = 0,\\ &\vec{A}.\vec{b} = 0, &&\vec{B}.\vec{b} = 2\pi, &&\vec{C}.\vec{b} = 0,\\ &\vec{A}.\vec{c} = 0, &&\vec{B}.\vec{c} = 0, &&\vec{C}.\vec{c} = 2\pi.\end{aligned}\right\} \quad \ldots (ii)$$

Any arbitrary set of vectors $\vec{a}, \vec{b}, \vec{c}$ of a given crystal lattice leads to the same set of reciprocal lattice points

$$\vec{G} = h\vec{A} + k\vec{B} + l\vec{C}, \quad \ldots (iii)$$

where h, k, l are integers. Any vector $\vec{G}$ of this form is called a 'reciprocal lattice vector'. It connects any two points of the reciprocal lattice.

The (direct), crystal lattice is a lattice in real or ordinary space ; the reciprocal lattice is a lattice in the so-called Fourier space.

The reciprocal lattice vector $\vec{G}$ and the (direct) crystal lattice vector $\vec{T}\left(= u\vec{a} + v\vec{b} + w\vec{c}\right)$ have the following relationship:

$$\begin{aligned}\vec{G}.\vec{T} &= \left(h\vec{A} + k\vec{B} + l\vec{C}\right).\left(u\vec{a} + v\vec{b} + w\vec{c}\right)\\ &= 2\pi\,(hu + kv + lw) \qquad \text{[from eq. } (ii)]\\ &= 2\pi N,\end{aligned}$$

where N is an integer. The quantity $hu + kv + lw$, being the sum of products of integers, is an integer. It follows that

$$e^{i\vec{G}.\vec{T}} = 1$$

because $$e^{i\,2\pi N} = 1$$

Properties of Reciprocal Lattice : Every crystal structure has two lattices associated with it, the (direct) crystal lattice' and the reciprocal lattice. The reciprocal lattice have the following properties with respect to the crystal lattice :

(i) *Every vector of the reciprocal lattice is normal to a set of lattice planes of the crystal lattice.*

Proof. A vector

$$\vec{G}(hkl) = h\vec{A} + k\vec{B} + l\vec{C}$$

will be normal to a plane through the three points $u\vec{a}, v\vec{b}, w\vec{c}$ of the crystal lattice if $\vec{G}$ is normal to any vector lying in that plane. The vectors $u\vec{a} - v\vec{b}, u\vec{a} - w\vec{c}$ and $v\vec{b} - w\vec{c}$ lie in the plane, and so we must have

or $$\left(h\vec{A} + k\vec{B} + l\vec{C}\right).\left(u\vec{a} - v\vec{b}\right) = 2\pi\ (hu - kv) = 0$$

or $$hu = kv.$$

Similarly, we must have

$$hu = lw,$$

and $$kv = lw,$$

These are satisfied, if

$$u = \frac{1}{h},\ v = \frac{1}{k},\ w = \frac{1}{l}.$$

But the Miller indices of the crystal; plane through the points $u\vec{a}, v\vec{b}, w\vec{c}$ are the $(h\ k\ l)$ of the last expression. Hence the vector $\vec{G}\,(hkl)$ of the reciprocal lattice is normal to the plane (hkl) of the crystal lattice.

(ii) *The magnitude of every reciprocal lattice vector is inversely proportional to the interplanar spacing of the corresponding set of lattice planes.*

Proof : Let $\vec{r}$ be any vector from the origin of the crystal lattice to a plane (hkl). Then the equation of the plane is*

$$\vec{r}\,.\hat{G}\,(hkl) = d(h\ k\ l)$$

where $\hat{G}\,(hkl)$ is the unit vector normal to the plane (hkl).

Let us consider one particular $\vec{r}$, say $\vec{r} = \vec{a}\,/h$. Then

$$\vec{r}\,.\hat{G}\,(hkl) = \frac{\vec{a}}{h}\cdot\left(h\vec{A} + k\vec{B} + l\vec{C}\right)$$

$$= \frac{1}{h}(2\pi h) = 2\pi$$

or $$\vec{r}\,.\hat{G}\,(hkl) = \frac{2\pi}{G(hkl)}\,. \qquad \left[\because \vec{G} = G\hat{G}\right]$$

Comparing it with the top expression, we get

$$\frac{2\pi}{G(hkl)} = d(hkl)$$

or $$G\ (hkl) \propto \frac{1}{d(hkl)}\,.$$

(iii) *The volume of a unit cell of the reciprocal lattice is inversely proportional to the volume of a unit cell of the crystal lattice.*

The reciprocal lattice cell is described by $\vec{A}, \vec{B}$ and $\vec{C}$. Its volume would be

$$\vec{A}\,.\vec{B}\times\vec{C} = (2\pi)^3\,\frac{\vec{b}\times\vec{c}}{\vec{a}\,.\vec{b}\times\vec{c}}\cdot\frac{(\vec{c}\times\vec{a})\times\left(\vec{a}\times\vec{b}\right)}{\left(\vec{a}\,.\vec{b}\times\vec{c}\right)^2}$$

$$= (2\pi)^3\,\frac{\left(\vec{b}\times\vec{c}\right).\left\{\vec{a}\left(\vec{c}\,.\vec{a}\times\vec{b}\right)\right\}}{\left(\vec{a}\,.\vec{b}\times\vec{c}\right)^3}$$

$$= (2\pi)^3 \frac{\left(\vec{b} \times \vec{c}\right) \cdot \left\{\vec{a}\left(\vec{a} \cdot \vec{b} \times \vec{c}\right)\right\}}{\left(\vec{a} \cdot \vec{b} \times \vec{c}\right)^3}$$

$$= \frac{(2\pi)^3}{\vec{a} \cdot \vec{b} \times \vec{c}}$$

where $\vec{a} \cdot \vec{b} \times \vec{c}$ is the volume of the unit cell of the crystal lattice.

Importance : A diffraction pattern of a crystal is a map of the reciprocal lattice of the crystal. The Bragg's diffraction law can be expressed in a simple way in terms of the reciprocal lattice vector $\vec{G}$ as

$$2\vec{k} \cdot \vec{G} + G^2 = 0$$

where $\vec{k}$ is the wave vector of the incident wave. This equation is very important in wave propagation and elastic scattering in crystal lattices. A similar result arises in the theory of the electron energy band structure of crystals. Thus the conception of reciprocal lattice is very important in the study of lattice dynamics of solids.

Brillouin Zones : *A Brillouin zone is defined as a Wigner-Seitz cell in the 'reciprocal lattice.* The sets of planes which are the perpendicular bisectors of the reciprocal lattice. Vectors are important in crystals, because a wave whose wave-vector drawn from the origin terminates on any of these planes satisfies Bragg's diffraction condition. These planes divide the Fourier space of the crystal into pieces. The central cell formed in this way in the reciprocal lattice is called the 'first Brillouin zone', and is of special importance in the theory of solids. *The first Brillouin zone is thus the smallest volume entirely enclosed by planes which are the perpendicular bisectors of the reciprocal lattice vectors drawn from the origin.*

The first Brillouin zone of an oblique lattice in two dimensions is constructed in Fig. below. We first draw an adequate number of vectors from O to nearby points in the reciprocal lattice. Next we draw lines perpendicular to these vectors at their midpoints. The *smallest* enclosed area is the first Brillouin zone. Any wave whose wave vector drawn from the origin terminates on the boundary of the zone will be diffracted by the crystal.

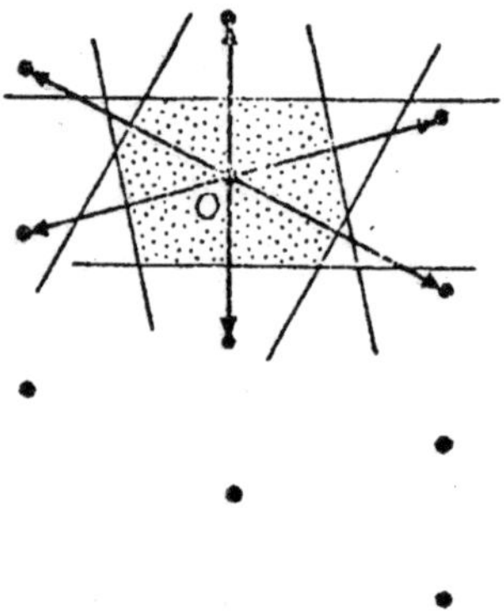

FIRST BRILLOUIN ZONE

Brillouin zones are an essential part of the analysis of the electronic energy-band structure of crystals.

Reciprocal Lattice to Simple Cubic (sc) Lattice : The translation vectors of a simple cubic lattice ($a = b = c$) may be written as

$$\vec{a} = a\hat{i}, \quad \vec{b} = a\hat{j}, \quad \vec{c} = a\hat{k}.$$

The volume of the unit cell is $\vec{a} \cdot \vec{b} \times \vec{c} = a^3$.

The translation vectors of the reciprocal lattice are found from the defining relations :

$$\left[\because \hat{j} \times \hat{k} = \hat{i}\right] \quad \left[\because \hat{j} \times \hat{k} = \hat{i}\right]$$

$$\left[\because \hat{k} \times \hat{i} = \hat{j}\right] \quad \left[\because \hat{k} \times \hat{i} = \hat{j}\right]$$

$$\left[\because \hat{i} \times \hat{j} = \hat{k}\right] \quad \left[\because \hat{i} \times \hat{j} = \hat{k}\right]$$

Thus the reciprocal lattice is itself a simple cubic lattice, now of ' lattice constant $2\pi/a$.

The boundaries of the first Brillouin zone are the planes normal to the six reciprocal lattice vectors $\pm \vec{A}$, $\pm \vec{B}$ and $\pm \vec{C}$ at their mid-points.

$$\left.\begin{aligned} \pm\frac{1}{2}\vec{A} &= \pm\frac{\pi}{a}\hat{i} \\ \pm\frac{1}{2}\vec{B} &= \pm\frac{\pi}{a}\hat{j} \\ \pm\frac{1}{2}\vec{C} &= \pm\frac{\pi}{a}\hat{k}. \end{aligned}\right\}$$

The six planes represented by these equations bound a cube of edge $2\pi/a$ and of volume $(2\pi/a)^3$. This "cube" is the first Brillouin zone of the simple cubic crystal lattice.

PROBLEMS

1. The glancing angle of the (110) plane of a simple cubic crystal (a = 2.814 Å) corresponding to second-order diffraction maximum for the X-rays of wavelength 0.710 Å is calculated as follows.

Solution : The distance between successive lattice planes defined by Miller indices h, k, l in a simple cubic lattice is given by

$$d_{hkl} = \frac{a}{\sqrt{(h^2 + k^2 + l^2)}}.$$

Here $a = 2.814$ Å, $h = 1$, $k = 1$, $l = 0$.

$$\therefore \qquad d_{110} = \frac{2.814 \text{ Å}}{\sqrt{2}} = 1.990 \text{ Å}$$

The Bragg's equation for X-ray diffraction is

$$2\, d_{110} \sin\theta = n\lambda.$$

This gives

$$\sin\theta = \frac{n\lambda}{2d_{110}}$$

Here n (order) = 2, λ = 0.710 Å

and $$d_{110} = 1.990 \text{ Å}.$$

$$\therefore \quad \sin\theta = \frac{2 \times 0.710}{2 \times 1.990} = 0.357$$

or $$\theta = \sin^{-1}(0.357) = 21°$$

2. The K_α line from the anticathode of an X-ray tube ... reflected in first order at an angle of 6° by a crystal of NaCl mounted on a Bragg spectrometer. In X. U. the wavelength of K_α radiation is calculated. Density of NaCl = 2170 kg/m^3, molecular weight of NaCl = 58.46 and Avogadro number N = 6.02 × 10^{26}/kg-molecule.

Solution : The interplanar spacing d of the crystal, in terms of molecular weight M and density ρ, is given by

$$d = \left(\frac{M}{2N\rho}\right)^{\frac{1}{3}}$$ (See prob. 7 last chapter)

Putting the given values :

$$d = \left[\frac{58.46 \text{ kg / kg - molecule}}{2 \times \left(6.02 \times 10^{26} \text{ / kg - molecule}\right) \times 2170 \, kg / m^3}\right]^{\frac{1}{3}}$$

$$= [22.37 \times 10^{-30} \text{ m}^3]^{1/3}$$

$$= 2.818 \times 10^{-10}\text{m}.$$

The Bragg's equation for specular reflection is

$$2d \sin\theta = n\lambda.$$

Here θ = 6° and n (order) = 1.

$$\therefore \quad \lambda = 2d \sin\theta = 2 \times (2.818 \times 10^{-10} \text{ m}) \times \sin 6°$$

$$= 2 \times (2.818 \times 10^{-10} \text{ m}) \times 0.1045$$

$$= 0.5890 \times 10^{-10} \text{m}$$

$$= 589.0 \times 10^{-13} \text{ m}$$

$$= 589.0 \text{ X.U.} \qquad (1 \text{ X.U.} = 10^{-13} \text{ m})$$

3. Electrons are accelerated to 344 volts and are reflected from a crystal. The first reflection maximum occurs when glancing angle is 60°. The interplanar spacing of the crystal is determined. Given: $h = 6.62 \times 10^{-34}$ joule-sec, $e = 1.6 \times 10^{-19}$ coulomb and electron mass $m = 9.1 \times 10^{-31}$ kg.

Solution : The wavelength λ associated with an electron of mass m moving with (non-relativistic) velocity v is given by

$$\lambda = \frac{h}{mv}.$$

If K be the kinetic energy of the electron, then

$$K = \tfrac{1}{2} mv^2 = eV,$$

where V is the accelerating voltage. From this

$$v = \sqrt{\left(\frac{2eV}{m}\right)}$$

and so
$$\lambda = \frac{h}{\sqrt{(2meV)}}$$

Substituting the given values :

$$\lambda = \frac{6.62 \times 10^{-34} \text{ joule-sec}}{\sqrt{\left[2 \times \left(9.1 \times 10^{-31} \text{kg}\right) \times \left(1.6 \times 10^{-19} \text{ coulomb}\right) \times \left(344 \text{ volt}\right)\right]}}$$

$$= 0.66 \times 10^{-10} \text{ meter} = 0.66 \text{ Å}.$$

Now, Bragg's equation for reflection is

$$2d \sin\theta = n\lambda.$$

For the 'first' reflection, $n = 1$ and $\theta = 60°$.

$$\therefore \quad 2d \sin 60° = \lambda = 0.66 \text{ Å}.$$

or
$$d = \frac{0.66 \text{ Å}}{2 \sin 60°}$$

$$= \frac{0.66 \text{ Å}}{2 \times 0.866} = 0.38 \text{ Å}$$

4. The angle of first order diffraction maximum when thermal neutrons (kinetic energy = 0.02 eV) undergo diffraction from a crystal plane (100) is calculated. The Bragg's spacing of the crystal is 2.0 Å.

The mass of neutron is 1.00898 amu and h = 6.62 × 10^{-34} joule-sec. Given : 1 eV = 1.6 × 10^{-19} joule and 1 amu = 1.66 × 10^{-27} kg.

Solution : The de Broglie wavelength of neutron, in usual notations, is

$$\lambda = \frac{h}{mv}$$

If K be the kinetic energy, then

$$K = \frac{1}{2} mv^2$$

and so
$$\lambda = \frac{h}{\sqrt{(2mK)}}.$$

Substituting the given values :

$$\lambda = \frac{6.62 \times 10^{-34} \text{ joule-sec}}{\sqrt{\left[2 \times \left(1.00898 \times 1.66 \times 10^{-27} \text{ kg}\right) \times \left(0.02 \times 1.6 \times 10^{-19} \text{ joule}\right)\right]}}$$

$$= 20 \times 10^{-10}\text{m} = 2.0 \text{ Å}.$$

The Bragg's equation for crystal diffraction is

$$2d \sin \theta = n\lambda.$$

Here $d = 2.0$ Å,

$n(\text{order}) = 1$

and $\lambda = 2.0$ Å.

$$\therefore \quad \sin \theta = \frac{n\lambda}{2d}$$

$$= \frac{1 \times 2.0\ \text{Å}}{2 \times 2.0\ \text{Å}} = 0.5$$

or $\theta = \sin^{-1}(0.5) = 30°$.

5. A two-dimensional lattice has the basis vectors $\vec{a} = 2\hat{i}$, $\vec{b} = \hat{i} + 2\hat{j}$. The basis vectors of the reciprocal lattice is found as follows.

Solution : Suppose the third vector $\vec{c}$ is parallel to the z-axis. Let $\vec{c} = \hat{k}$.

The vectors of the reciprocal lattice are

$$\vec{A} = 2\pi \frac{\vec{b} \times \vec{c}}{\vec{a} . \vec{b} \times \vec{c}}$$

$$\vec{B} = 2\pi \frac{\vec{c} \times \vec{a}}{\vec{a} . \vec{b} \times \vec{c}}.$$

Now,

$$\vec{b} \times \vec{c} = (\hat{i} + 2\hat{j}) \times \hat{k}$$

$$= \hat{i} \times \hat{k} + 2\hat{j} \times \hat{k} \text{ (vector product is distributive)}$$

$$= -\hat{j} + 2\hat{i}, \qquad \text{(rule of vector product)}$$

$$\vec{c} \times \vec{a} = \hat{k} \times 2\hat{i} = 2\hat{j},$$

and $\vec{a}.\vec{b}\times\vec{c} = \begin{vmatrix} a_x & a_y & a_z \\ b_x & b_y & b_z \\ c_x & c_y & c_z \end{vmatrix}$

$$= \begin{vmatrix} 2 & 0 & 0 \\ 1 & 2 & 0 \\ 0 & 0 & 1 \end{vmatrix} = 4.$$

Making these substitutions, we get

$$\vec{A} = \pi\hat{i} - \frac{1}{2}\pi\hat{j}\ ;\ \vec{B} = \pi\hat{j}$$

The crystal lattice vectors $\vec{a}, \vec{b}$; and the reciprocal lattice vectors $\vec{A}, \vec{B}$ have been sketched in the Fig. $\vec{A}$ and $\vec{B}$ are perpendicular to sets of planes in the crystal lattice, namely, to the lines parallel to $\vec{b}$ and $\vec{a}$ respectively.

CRYSTAL LATTICE

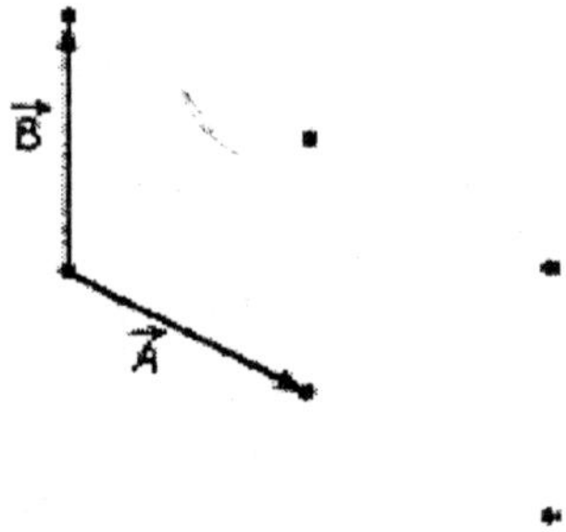

RECIPROCAL LATTICE

9

Space Lattice

Space Lattice in Crystal Structure : A crystal (such as rock-salt, quartz, calcite, diamond, mica, etc.) is a solid composed of *geometrically-regular* arrangement of atoms (or ions or molecules) in space. When struck into pieces, all pieces have the same shape with same angles between corresponding faces. These angles are characteristic of the given crystal.

Bravais assumed that a crystal is made up of a three-dimensional array of points such that each point is surrounded by the neighbouring points in an *identical* way. Such an array of points is known as 'Bravais lattice' or 'space lattice'. Thus, *a lattice is a regular arrangement of points extended repeatedly in space.*

Translation Vectors

Mathematically, a lattice is expressed in terms of three translation vectors $\vec{a}, \vec{b}, \vec{c}$ such that the atomic arrangement looks exactly identical when viewed from points $\vec{r}$ and $\vec{r}\,'$ (Fig.), where

$$\vec{r}\,' = \vec{r} + u\vec{a} + v\vec{b} + w\vec{c}\,.$$

u, v, w are arbitrary integers.

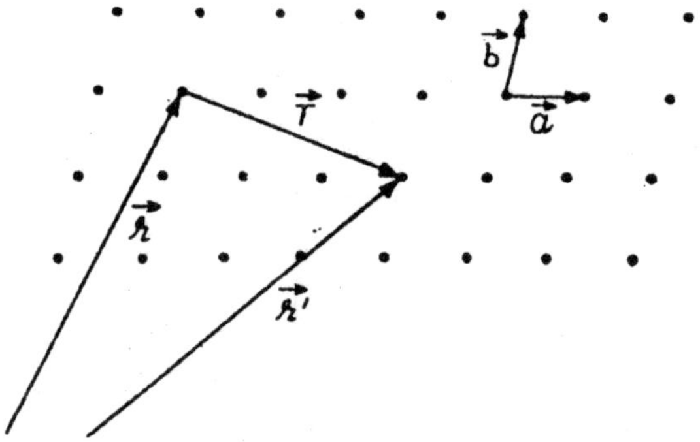

Basis : In the simplest crystals (such as copper, silver, sodium, etc) there is a single atom (or ion) at each lattice point. More often, however, there is a group of several atoms (or ions) attach lattice point. This group is called the 'basis'. Each basis is identical in composition, arrangement and orientation with any other basis.

The crystal structure is formed when a basis of atoms is attached identically to each lattice point, as shown in Fig.

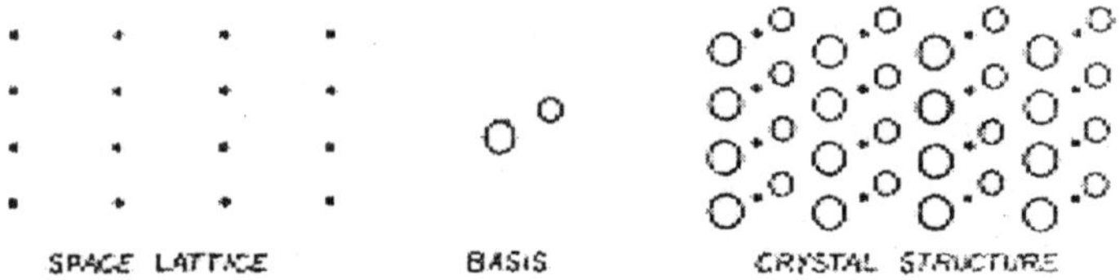

Obviously, the logical relation is

lattice + basis = crystal structure.

There is consequently a distinction between the structure of a crystal, which is the actual ordering in space of its constituent atoms, and the corresponding crystal lattice, which is a geometrical abstraction useful for classifying the crystal.

The translation vectors $\vec{a}, \vec{b}, \vec{c}$ of the lattice define the crystal axes.

A 'lattice translation operation' is defined as the displacement of a crystal parallel to itself by a crystal translation vector

$$\vec{T} = u\vec{a} + v\vec{b} + w\vec{c}$$

where the vector $\vec{T}$ connects any two lattice points (Fig.).

Unit Cell : The parallelepiped formed by the translation vectors $\vec{a}, \vec{b}, \vec{c}$ as edges is called a 'unit cell' of the space lattice. The angles between $(\vec{b}, \vec{c})$, $(\vec{c}, \vec{a})$ and $(\vec{a}, \vec{b})$ are denoted by α, β and γ respectively (Fig.).

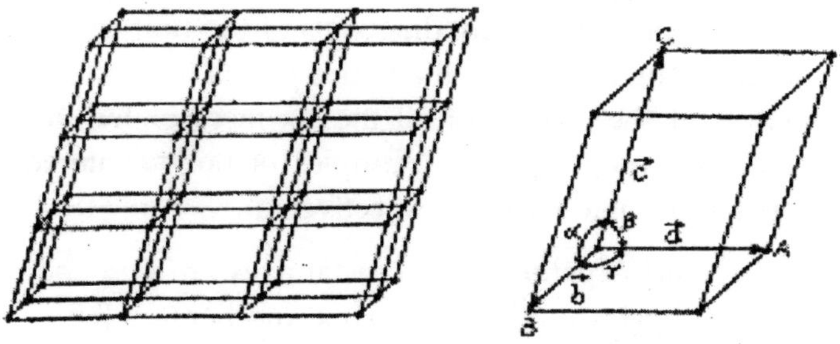

SPACE LATTICE (a) UNIT CELL (b)

Fig. shows part of a space lattice. It is possible to isolate a unit cell, and the space lattice can be constructed by repeatedly translating the unit cell along its edges.

Seven Crystal Systems

There are seven types of crystals depending upon their axial ratios ($a : b : c$) and angles between them (α, β, γ). Bravais showed that they can give rise to 14 types of space lattices. These crystals and the corresponding Bravais lattices with their characteristic features are as below :

Cubic Crystals : In cubic crystals, the crystal axes are perpendicular to one another (α = β = γ = 90°) and the repetitive interval is the same along the three axes ($a = b = c$). Cubic lattices may be simple, body-centered or face-centered as shown in below Fig.

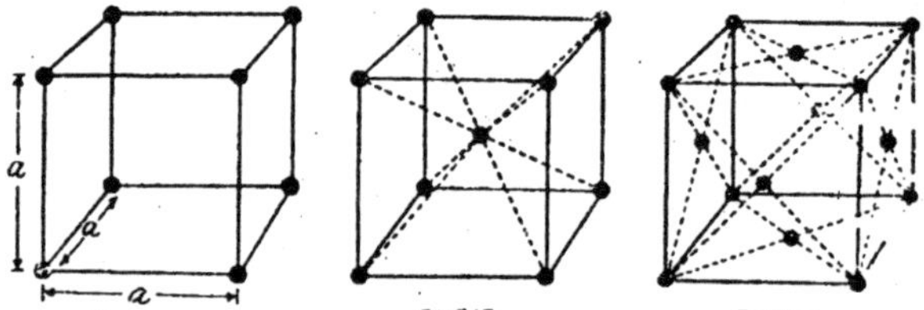

CUBIC (SIMPLE) (a) — CUBIC (BODY-CENTERED) (b) — CUBIC (FACE-CENTERED) (c)

In the simple cubic lattice, the lattice points are situated only at the corners of the unit cell (Fig *a*). In the body-centered lattice the lattice points are situated at the corners and also at the intersection of the body diagonals of the unit cell (Fig. *b*). In the face-centered lattice, the lattice points lie at the corners as well as at the centers of all the six faces of the unit cell (Fig. *c*). CsCl and NaCl are examples of body-centered and face-centered cubic lattices respectively.

Tetragonal Crystals : The crystal axes are perpendicular Solid State Physics and Devices to one another ($\alpha = \beta = \gamma = 90°$). The repetitive intervals along two axes are the same, but the interval along the third axis is different ($a = b \neq c$). Tetragonal lattices may be simple or body-centered.

Orthorhombic Crystals : The crystal axes are perpendicular to one another ($\alpha = \beta = \gamma = 90°$), but the repetitive intervals are different along all three axes ($a \neq b \neq c$). Orthorhombic crystals may be simple, base-centered, body-centered, or face-centered.

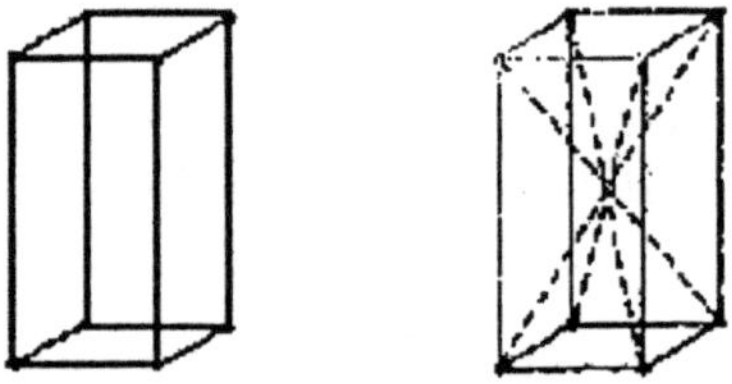

SIMPLE — BODY-CENTERED

TETRAGONAL

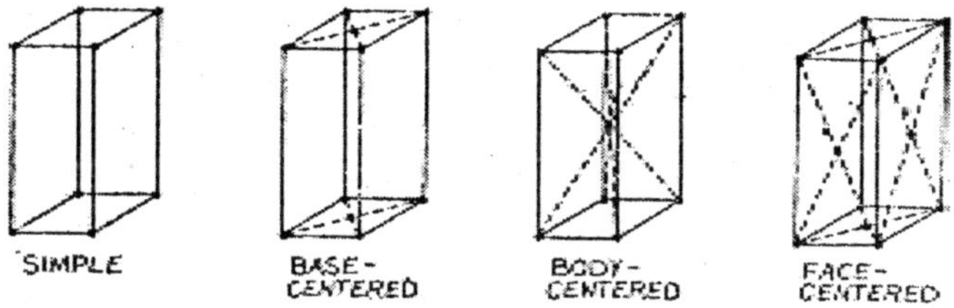

ORTHORHOMBIC

Monoclinic Crystals: Two of the crystal axes are not perpendicular to each other, but the third is perpendicular to both of them ($\alpha = \gamma = 90° \neq \beta$). The repetitive intervals are different along all three axes ($a \neq b \neq c$). Monoclinic lattices may be simple or base-centered (Fig.).

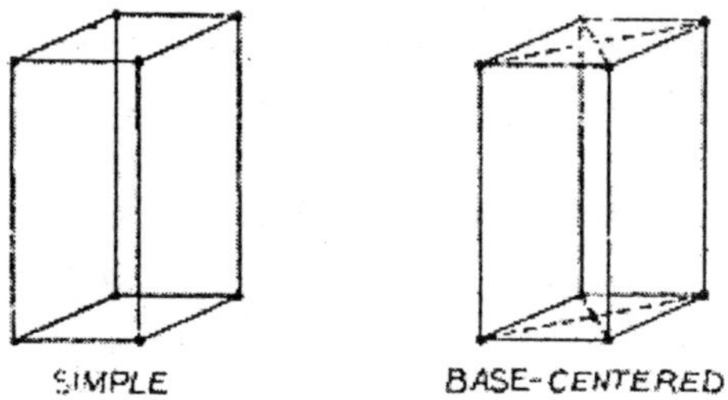

MONOCLINIC

Triclinic Crystals : None of the crystal axes is perpendicular to any of the others ($\alpha \neq \beta \neq \gamma$), and the repetitive intervals are different along all three axes ($a \neq b \neq c$). The triclinic lattice is only simple.

Trigonal (or Rhombohedral) Crystals : The angles between each pair of crystal axes are the same but different from 90° ($\alpha = \beta = \gamma \neq 90°$). The repetitive interval is the same along all three axes ($a = b = c$). The trigonal lattice is only simple (Fig.).

Hexagonal Crystals : Two of the crystal axes are 60° apart while the third is perpendicular to both of them ($\alpha = \beta = 90°$, $\gamma = 120°$). The repetitive intervals are the same along the two 60°-apart axes, but the interval along the third is different ($a = b \neq c$). The hexagonal lattice is only simple (Fig.).

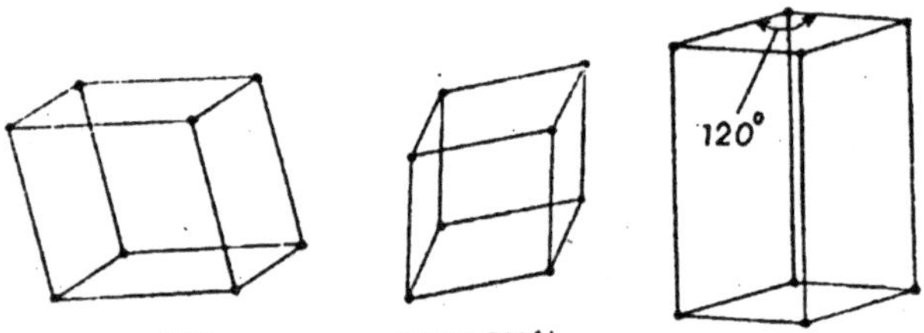

TRICLINIC TRIGONAL HEXAGONAL

The above description can be tabulated as below :

Crystal class	*Intercepts on Axes*	*Angles between Axes*	*Types of Bravias space lattices*
Cubic	$a = b = c$	$\alpha = \beta = \gamma = 90°$	simple, body-centered, face-centered
Tetragonal	$a = b \neq c$	$\alpha = \beta = \gamma = 90°$	simple, body-centered
Orthorhombic	$a \neq b \neq c$	$\alpha = \beta = \gamma = 90°$	simple, base-centered, body-centered, face centered
Monoclinic	$a \neq b \neq c$	$\alpha = \gamma = 90° \neq \beta$	simple, base-centered
Triclinic	$a \neq b \neq c$	$\alpha \neq \beta \neq \gamma$	simple
Trigonal	$a = b = c$	$\alpha = \beta = \gamma \neq 90°$	simple
Hexagonal	$a = b \neq c$	$\alpha = \beta = 90°, \gamma = 120°$	simple

Cubic Crystal Lattices : The type of crystal lattice that is equivalent in all directions in space is the "cubic" lattice ($a = b = c$; $\alpha = \beta = \gamma = 90°$). There are three types of cubic lattices ; simple, body-centered and face-centered. Let us consider them in detail:

Simple Cubic (sc) Lattice : In this space lattice, the lattice points are situated only at the corners of the unit cells constituting the three-dimensional structure (Fig.). Each cell has eight corners, and eight cells meet at each corner. Thus only one-eight of a lattice point belongs to each cell. That is, *there is only 1 lattice point (or 1 atom) per unit cell.* A unit cell containing only 1 lattice point is called a 'primitive' cell. Since the simple cubic lattice is built of primitive cells, it is also known as cubic P lattice.

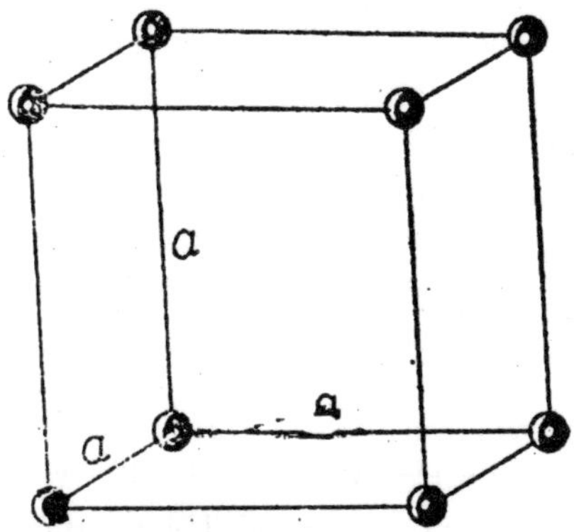

CUBIC P
(SIMPLE)

Coordination Number : *The coordination number is defined as the number of nearest neighbours around any lattice point (or atom) in the crystal lattice.* Let us compute it for the simple cubic lattice (Fig.).

Taking any one of the lattice points as origin, and the three edges passing through that point as x, y and z axes, the positions of the *nearest* neighbours of the origin are

$$\pm a\hat{i}, \pm a\hat{j}, \pm a\hat{k}$$

where $\hat{i}$, $\hat{j}$, $\hat{k}$ are unit vectors along the x, y, z axis respectively. The coordinates of these points (or atoms) nearest the origin are

$$(\pm a, 0, 0) ; (0, \pm a, 0) ; (0, 0, \pm a).$$

Their number is obviously 6. Hence *the coordination number of a simple cubic lattice is 6.* The distance between two nearest neighbours is a, which is the 'lattice constant' (length of each edge of the unit cell).

Body-centered Cubic (bcc) Lattice : In this lattice, the lattice points (or atoms) are situated at each corner of the unit cell and also at the Intersection of the *body* diagonals of the cell (Fig.). This lattice is also known as cubic I lattice.

Each cell has 8 corners and 8 cells meet at each corner. Thus only 1/8th of a lattice point (or atom) at a corner belongs to any one cell. Also there is a lattice point (or atom) at the body centre of each cell. Thus the total number of lattice points in any one cell is $\left(\frac{1}{8}\times 8\right)$ + 1=2, that is, *a bcc lattice has 2 lattice points (or atoms) per unit cell.*

To compute the coordination number, let us take the lattice point at the body centre as the origin; the x, y, z axes being parallel to the edges of the unit cell. The positions of the *nearest* neighbours of the origin are

$$\left(\pm\frac{a}{2}\hat{i},\pm\frac{a}{2}\hat{j},\pm\frac{a}{2}\hat{k}\right)$$

The coordinates of these points (or atoms) nearest the origin are

$$\left(\frac{a}{2},\frac{a}{2},\frac{a}{2}\right);\left(-\frac{a}{2},\frac{a}{2},\frac{a}{2}\right);\left(\frac{a}{2},-\frac{a}{2},\frac{a}{2}\right);$$

$$\left(\frac{a}{2},\frac{a}{2},-\frac{a}{2}\right);\left(-\frac{a}{2},-\frac{a}{2},\frac{a}{2}\right);$$

$$\left(-\frac{a}{2},\frac{a}{2},-\frac{a}{2}\right);\left(\frac{a}{2},-\frac{a}{2},-\frac{a}{2}\right);$$

$$\left(-\frac{a}{2},-\frac{a}{2},-\frac{a}{2}\right).$$

Their number is obviously 8. Hence *the coordination number of a bcc lattice is 8.*

The distance between any two nearest neighbours is

$$\sqrt{\left[\left(\frac{a}{2}\right)^2+\left(\frac{a}{2}\right)^2+\left(\frac{a}{2}\right)^2\right]}=\frac{\sqrt{3}}{2}a$$

A typical bcc structure is CsCl, and so the bcc lattice is often known as the CsCl structure. In this structure each ion is surrounded by 8 neighbours of the opposite charge (Fig.).

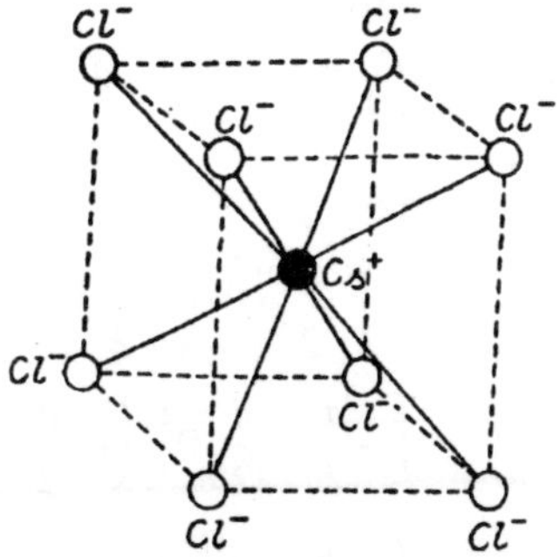

Face-centered Cubic (fcc) Lattice : In this lattice, the lattice points (or atoms) are situated at all the eight corners of the unit cell and also at the centers of all the six faces of the cell. This lattice is also known as cubic *F* lattice.

Each unit cell has 8 corners and 8 cells meet at each corner. Thus only 1/8th of a lattice point (or atom) at a corner belongs to any one cell. Similarly, a lattice point at the center of a face of the cell is shared by 2 cells, that is, only ½ of lattice point belongs to any one cell. Since a cell has 8 corners and 6 faces, the total number of lattice points belonging to any one cell is

$\left(\frac{1}{8}\times 8\right)+\left(\frac{1}{2}\times 6\right) = 4$, that is, *an fcc lattice has 4 lattice points (or atoms) per unit cell.*

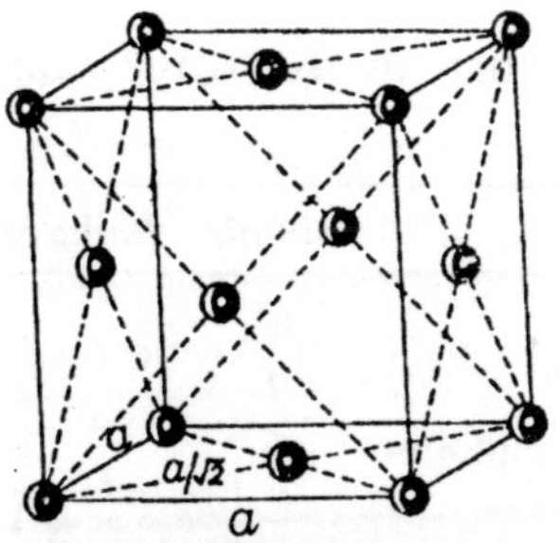

CUBIC F
(FACE-CENTERED)

To compute the coordination number, let us take any of the lattice points as the origin ; the x, y, z axes being parallel to the edges of the unit cell. The positions of the *nearest* neighbours of the origin are

$$\left(\pm\frac{a}{2}\hat{i}\pm\frac{a}{2}\hat{j}, \pm\frac{a}{2}\hat{j}\pm\frac{a}{2}\hat{k}, \pm\frac{a}{2}\hat{k}\pm\frac{a}{2}\hat{i}\right)$$

The coordinates of these points (or atoms) are

$$\left(\frac{a}{2},\frac{a}{2},0\right);\left(-\frac{a}{2},\frac{a}{2},0\right);\left(\frac{a}{2},-\frac{a}{2},0\right);$$

$$\left(-\frac{a}{2},-\frac{a}{2},0\right);\left(0,\frac{a}{2},\frac{a}{2}\right);\left(0,-\frac{a}{2},\frac{a}{2}\right);$$

$$\left(0,\frac{a}{2},-\frac{a}{2}\right);\left(0,-\frac{a}{2},-\frac{a}{2}\right);\left(\frac{a}{2},0,\frac{a}{2}\right);$$

$$\left(-\frac{a}{2},0,\frac{a}{2}\right);\left(\frac{a}{2},0,-\frac{a}{2}\right);\left(-\frac{a}{2},0,-\frac{a}{2}\right).$$

There number is obviously 12. Hence *the coordination number of an fcc lattice is 12.*

The distance between two nearest neighbours is

$$\sqrt{\left[\left(\frac{a}{2}\right)^2+\left(\frac{a}{2}\right)^2+(0)^2\right]} = \frac{1}{\sqrt{2}}a.$$

This lattice gives a very efficient packing (more atoms per unit volume) and so it is usually the most stable structure. The most familiar fcc lattice is NaCl, and so the fcc lattice is often called NaCl structure.

The characteristics of the three cubic lattices may be tabulated as below :

	Simple	*Body-centered*	*Face-centered*
Volume of unit cell	a^3	a^3	a^3
Lattice points per cell	1	2	4
Number of nearest neighbours	6	8	12
Nearest-neighbour distance	a	$\frac{\sqrt{3}}{2}a$	$\frac{1}{\sqrt{2}}a$

NaCl Structure : The NaCl crystal is a system of Na^+ and Cl^- ions arranged alternately in a 'cubic' pattern in space so that the electrostatic attraction between the nearest neighbours is maximum. Each ion lies on three rows of equally-spaced ions at right angles to one another.

In Fig. is shown a unit cell of NaCl lattice. The Na^+ ions are situated at the corners as well as at the centres of the faces of the cube, that is, Na^+ ions lie on a fcc lattice. So do the Cl^- ions, their lattice being relatively displaced half the edge of the unit cell along each axis. Thus NaCl crystal can be thought of as composed of interleaved fcc Na^+ and Cl^- sublattices.

Each cell has 8 corners and 8 cells meet at each corner. Thus an ion at a corner of the cell is shared by 8 cells, *i.e.*, only $\frac{1}{8}$ ion belongs to any one cell. Similarly, an ion at the centre of a face of the cell is shared by 2 cells, *i.e.* only ½ ion belongs to any one cell. Since a cell has 8 corners and 6 faces, it has

$\left(\frac{1}{8} \times 8\right) + \left(\frac{1}{2} \times 6\right)$ = 4 ions of one kind, and similarly 4 ions of the other kind. Thus *there are* 4 $Na^+ - Cl^-$ *ion pairs per unit cell.*

Each Na^+ ion has 6 Cl^- ions as nearest neighbours, and similarly each Cl^- ion has 6 Na^+ ions. Hence the coordination number of NaCl is 6, the same as that for simple cubic lattice.

Since a given Na^+ ion is attracted by 6 nearest Cl^- neighbours, and does not belong to any single Cl^- ion, the NaCl crystal cannot be thought of as being composed of molecules.

Unit Cell and Lattice Constant of a Space Lattice : A unit cell is a parallelopiped isolated from a space lattice, having its edges along the translation vectors. The space lattice can be constructed by repeatedly translating the unit cell along its edges.

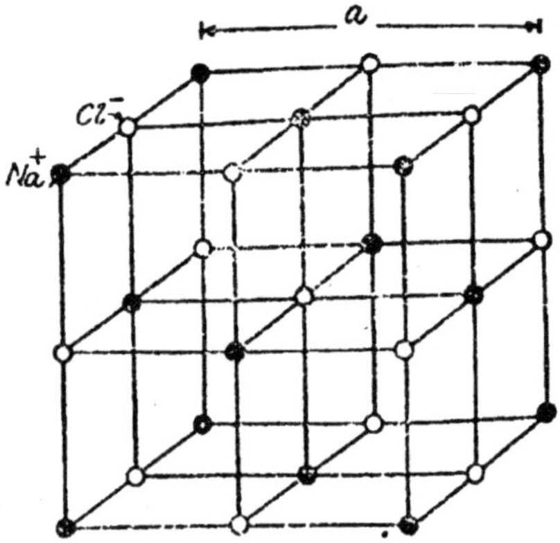

The lattice constant of a cubic lattice is the dimension of its unit cell. In the figure above, the lattice constant is a.

Let us consider a cubic crystal lattice of lattice constant a. The volume of the unit cell is a^3. If ρ be the density of the cell, then

$$\rho = \frac{m}{a^3} \qquad \text{...}(i)$$

where m is the mass per unit cell.

Let n be the number of molecules per unit cell. The mass of one molecule is M/N, where M is the molecular weight of the crystal and N is Avogadro's number. Then, the mass per unit cell is

$$m = n\frac{M}{N}$$

Substituting this value of m in eq. (i), we get

$$\rho = \frac{nM}{Na^3}$$

or
$$a = \left(\frac{nM}{N\rho}\right)^{\frac{1}{3}}.$$

From this relation the value of the lattice constant can be calculated.

Lattice Planes of a Crystal : A crystal lattice may be considered as an aggregate of a set of parallel, equally-spaced planes passing through the lattice points. The planes are called 'lattice planes', and the perpendicular distance between adjacent planes is called 'interplanar spacing'.

A given space lattice may have an infinite sets of lattice planes, each having its characteristic interplanar spacing. In below given figure are shown three sets of lattice planes with interplanar spacings d_1, d_2 and d_3. Each set accommodates *all* the lattice points. Out of these, only those which have high density of lattice points are significant and show diffraction of X-rays. They are known as 'Bragg planes' or 'Cleavage planes'. When a crystal is struck, it breaks most easily across its cleanvage planes.

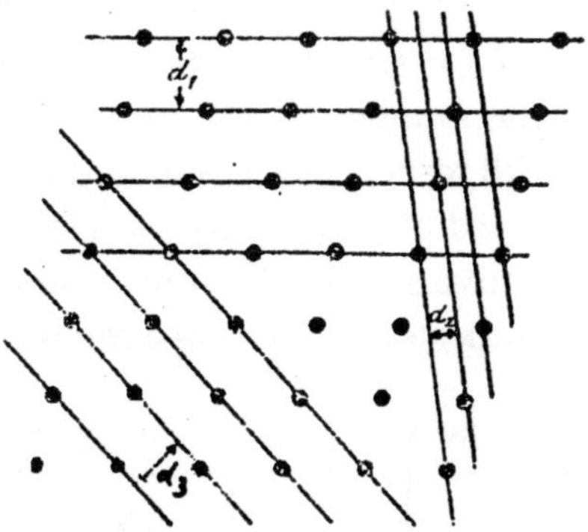

Position and Orientation of Lattice Planes in a Crystal—Miller Indices : *The position and orientation of a lattice plane in a crystal are determined by three smallest whole numbers which have same ratios with one another as the reciprocals of the intercepts of the plane on the three crystal axes.* These numbers, denoted by *h, k, l,* are known as the 'Miller indices' of that plane or of any plane parallel to it); and the plane is specified as (*hkl*).

Let *OX, OY, OZ* be three axes parallel to the crystal axes the planes *XOY, XOZ* and *ZOY* are parallel to the faces of the crystal. Besides these, other possible cleavage planes also exist.

Let ABC be a standard plane cutting all the three axes at A, B and C with intercepts OA, OB and $OC = a, b$ and c respectively, where $\vec{a}, \vec{b}$ and $\vec{c}$ are the translation vectors of the crystal lattice.

The directions of the other faces of the crystal are governed by the 'law of rational indices'. This law states that *a face which is parallel to a plane whose intercepts on the three axes are m_1a, m_2b and m_3c where m_1, m_2 and m_3 are small whole numbers, is a possible face of the crystal.* Thus, if $A'B'C'$ be a possible crystal face, then

$$OA' : OB' : OC' = m_1a : m_2b : m_3c$$

$$= \frac{a}{m_2m_3} : \frac{b}{m_1m_3} : \frac{c}{m_1m_2}$$

$$= \frac{a}{h} : \frac{b}{k} : \frac{c}{l}$$

where h $(= m_2\,m_3)$, k $(= m_1m_3)$ and $l(= m_1m_2)$ are again small whole numbers. The numbers h, k, l are the Miller indices of the plane $A'B'C'$ with respect to the standard plane. The Miller indices of the standard plane are always (111). If a plane cuts an axis on the negative side of the origin, the corresponding index is negative, indicated by placing a minus sign above the index, as $(h, \bar{k}, l)$.

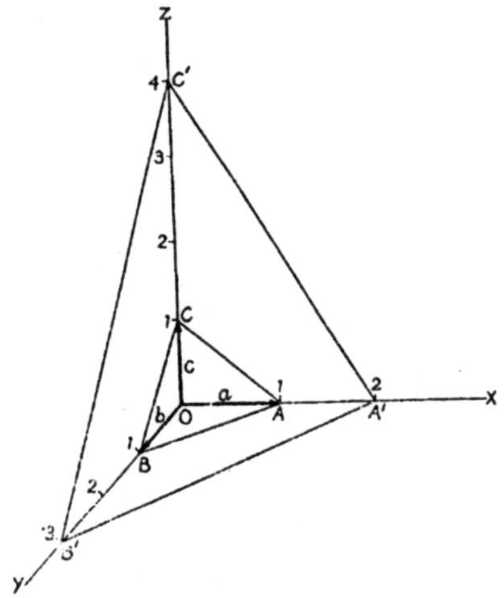

As an example, suppose a plane cuts the X-axis at $2a$, the Y-axis at $3b$ and the Z-axis at $4c$. Then, from the law of rational indices, we have

$$2a : 3b : 4c = \frac{a}{h} : \frac{b}{k} : \frac{c}{l}$$

or
$$h : k : l = \frac{1}{2} : \frac{1}{3} : \frac{1}{4}$$

The smallest whole numbers which have same ratios as the reciprocals $\frac{1}{2}, \frac{1}{3}$ and $\frac{1}{4}$ are 6 : 4 : 3. Hence the miller indices of the plane are h = 6, k = 4 and l= 3; or the plane is (643).

Directional Indices : The indices of a direction in a crystal are the set of the smallest integers which have the same ratios as the components of a vector in the desired direction, referred to the axes. The directional indices (integers) are written as [*uvw*]. Thus the X-axis is the [100] direction, the —Y-axis is the $\left[01\bar{0}\right]$ direction, and so on.

In cubic crystals, the normal to a plane (*hkl*) is the direction [*hkl*].

Sketching of Lattice Planes in a Cubic Crystal : Let us consider a simple cubic lattice plane ($\alpha = \beta = \gamma = 90°$ and $a = b = c$) The axes *OX, OY, OZ* form a right-angled set . The face *ABFE* in Fig (*a*), or any plane parallel to it, has an intercept on X-axis, but it is parallel to the Y and Z-axis *i.e.* the intercepts on the Y and Z-axis are each infinity.

If the intercept on the X-axis is taken as 1, then the miller indices are the reciprocals of 1, ∞, ∞, *i.e.*, 1, 0, 0; and the plane is (100). The plane parallel to *ABFE* but intercepting the X-axis on the negative side of the origin is $\left(\bar{1}00\right)$.

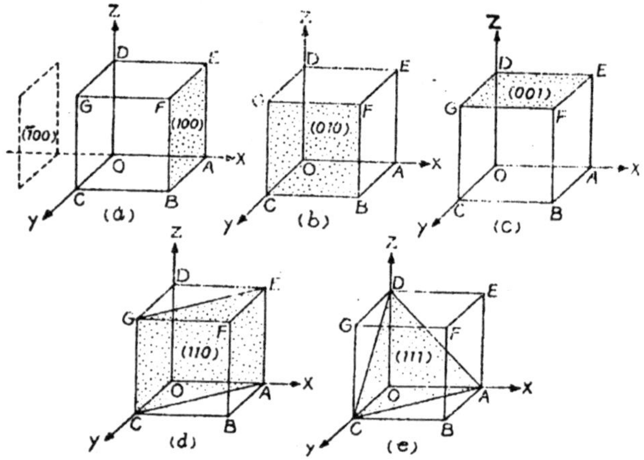

Similarly a plane parallel to *BCGF* is a (010) plane (Fig. *b*) and the one parallel to *DEFG* is a (001) plane (Fig. *c*).

The diagonal plane *ACGE* (Fig. *d*) has equal intercepts of 1 on *X* and *Y* axes and is parallel to the *Z*-axis. It is therefore a (110) plane.

The plane *ACD* (Fig. *e*) makes equal intercepts on the three axes and so its miller indices are 1, 1, 1 and so it is a (111) plane.

Interplanar Spacing : The separation between successive lattice planes of cubic, tetragonal and orthorhombic crystals, for which $\alpha = \beta = \gamma = 90°$, can be deduced as follows.

Let *OX*, *OY* and *OZ* be three axes parallel to the crystal axes (Fig.). Let *ABC* be one of a series of parallel lattice planes in the crystal. Let the plane *ABC* have intercepts $OA = a/h$, $OB = b/k$ and $OC = c/l$. Let us suppose that the origin *O* lies in the plane *adjacent* to *ABC*. Then, *ON*, the length of the normal from the origin to the plane *ABC*, is equal to the interplanar distance*d*. Let θ_a, θ_b and θ_c be the angles which *ON* makes with the three crystallographic axes respectively. Then the direction-cosines of *ON* are

$$\cos\theta_a = \frac{ON}{OA}, \cos\theta_b = \frac{ON}{OB} \text{ and } \cos\theta_c = \frac{ON}{OC}.$$

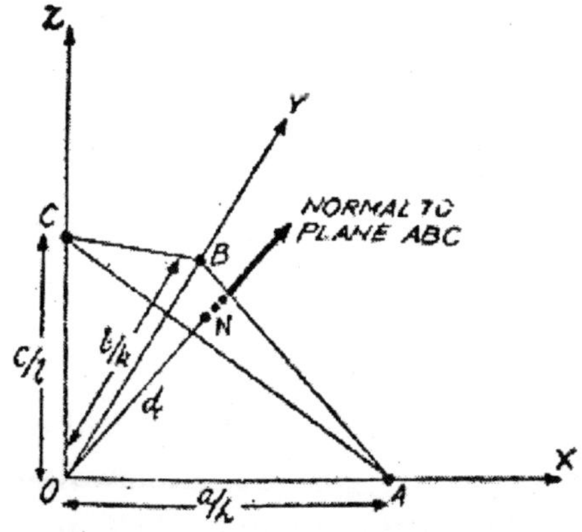

We know that the sum of the squares of the direction-cosines of a line is equal to unity ($\cos^2\theta_a + \cos^2\theta_b + \cos^2\theta_c = 1$). Therefore

$$\left(\frac{ON}{OA}\right)^2 + \left(\frac{ON}{OB}\right)^2 + \left(\frac{ON}{OC}\right)^2 = 1$$

or
$$\left(\frac{d}{a/h}\right)^2 + \left(\frac{d}{b/k}\right)^2 + \left(\frac{d}{c/l}\right)^2 = 1$$

or
$$d^2\left[\frac{h^2}{a^2} + \frac{k^2}{b^2} + \frac{l^2}{c^2}\right] = 1$$

or
$$d = \frac{1}{\sqrt{\left(\frac{h^2}{a^2} + \frac{k^2}{b^2} + \frac{l^2}{c^2}\right)}}$$

For a cubic crystal, the lengths of the sides of a unit cell are equal *i.e.* $a = b = c$.

$\therefore$
$$\boxed{d = \frac{a}{\sqrt{(h^2 + k^2 + l^2)}}}$$

Let us calculate the values of d for some of the series of lattice planes in a *simple* cubic crystal. For the (100) planes, we have $h = 1$, $k = 0$, and $l = 0$

$\therefore \qquad d_{100} = a.$

For the (110) planes, we have $h = 1$, $k = 1$, and $l = 0$

$$\therefore \qquad d_{110} = a\sqrt{2}\cdot$$

For the (111) planes, we have $h = 1$, $k = 1$, and $l = 1$.

$$\therefore \qquad d_{111} = a\sqrt{3}\cdot$$

$$\text{Thus} d_{100} : d_{110} : d_{111} = 1 : \frac{1}{\sqrt{2}} : \frac{1}{\sqrt{3}} = 1 : 0.71 : 0.58.$$

Spacing between Lattice Planes in BCC and FCC Crystals: Fig. (*a*) and (*b*) show bcc lattices. Comparing with simple cubic lattice, it is seen that in bcc lattice there exist additional planes half-way between the (100) planes and also between the (111) planes.

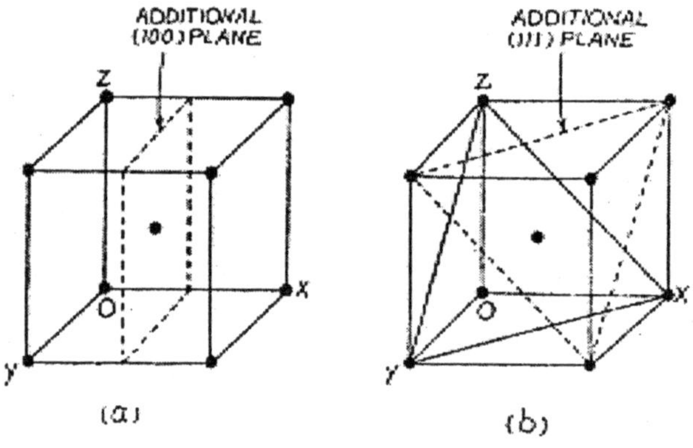

Therefore, in the light of spacings in the simple cubic lattice ($d_{100} = a$, $d_{110} = a/\sqrt{2}$, $d_{111} = a/\sqrt{3}$), the spacings between (100), (110) and (111) planes in the bcc lattice are given by

$$d_{100} = \frac{a}{2}$$

$$d_{110} = \frac{a}{\sqrt{2}}$$

and $$d_{111} = \frac{a}{2\sqrt{3}}.$$

Thus $$d_{100} : d_{110} : d_{111} = 1 : \sqrt{2} : \frac{1}{\sqrt{3}}$$

Fig. (*a*) and (*b*) show fcc lattices. Again comparing with simple cubic lattice, in fee lattice there exist additional planes halfway between the (100) planes and also between the (110) planes.

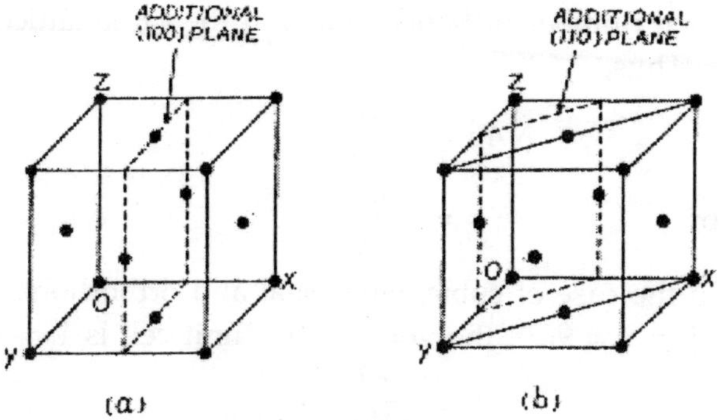

Therefore, in the light of the spacings in the simple cubic lattice, the spacings between (100), (110) and (111) planes in the fee lattice are given by

$$d_{100} = \frac{a}{2}$$

$$d_{110} = \frac{a}{2\sqrt{2}}$$

and $$d_{111} = \frac{a}{\sqrt{3}}$$

Thus, $$d_{100} : d_{110} : d_{111} = 1 : \frac{1}{\sqrt{2}} : \frac{2}{\sqrt{3}}.$$

Density of Lattice Points in a Lattice Plane : Let us consider *N* parallel lattice planes in a crystal lattice. Let *A* be the cross-sectional area of each lattice plane and *d* the spacing between successive planes. The volume of the lattice space under consideration is then NAd. If V be the volume of a unit cell, then the number of unit cells in the lattice space is $\frac{NAd}{V}$.

If n be the number of lattice points per unit cell, then the total number of lattice points in the lattice space is $\frac{nNAd}{V}$.

Now, let ρ be the density of lattice points in a lattice plane, *i.e.*, the number of lattice points per unit area of the plane. Then the total number of lattice points in the same lattice space is *NAP*. Thus,

$$NA\rho = \frac{nNAd}{V}$$

or

$$\rho = \frac{nd}{V}$$

In the case of cubic, tetragonal and orthorhombic lattices ($\alpha = \beta = \gamma = 90°$), the volume of a unit cell is $V = abc$. Thus

$$\rho = \frac{nd}{abc}.$$

For primitive lattice in each of these systems, there is one lattice point per unit cell, *i.e.*, $n = 1$. In such cases

$$\rho = \frac{d}{abc}$$

Hexagonal Close-packed Structure—(HCP) : A crystal whose constituents atoms are so arranged as to occupy the least possible volume is said to have a 'close-packed' structure. Such structures occur when the bonding forces are spherically symmetric (as in inert gases) or very nearly so (as in metals).

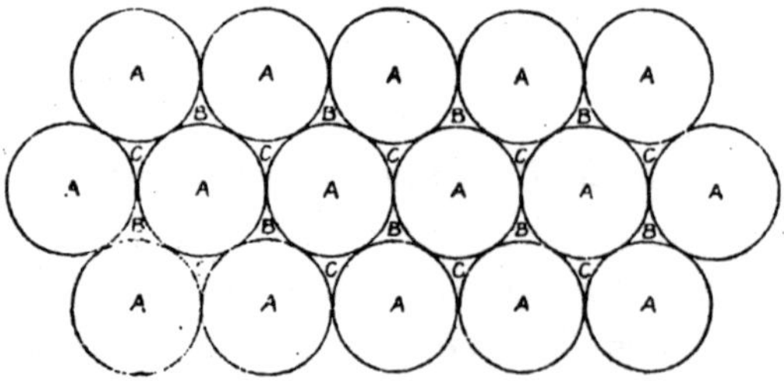

To understand the close-packed structure, let us consider identical spheres. The spheres can be arranged in a single *closest-packed* layer by placing each sphere in contact with six others (Fig.). The layer then assumes hexagonal shape. A second identical layer can be formed over this by placing spheres on the hollows *B*, each formed by three spheres in the bottom layer. Now, a third layer can be added in two different ways. In one, the spheres in the third layer are placed over the hollows *C* ; the resulting three-dimensional structure is then 'face-centered cubic' (or 'cubic close-packed') structure. Thus the packing in the fcc structure is *ABC ABC ABC*.........

Alternatively, the spheres in the third layer can be placed over *A*'s, that is, directly over the spheres in the first layer ; the resulting three-dimensional structure is then 'hexagonal closed-packed' (hcp). Thus the packing in the hcp structure is *AB AB AB*.........

In the hcp structure, the spheres in *alternate* layers are directly above one another (Fig.). The atom positions in this structure do not constitute a space lattice. The space lattice is simple hexagonal with a basis of two identical atoms associated with each lattice point.

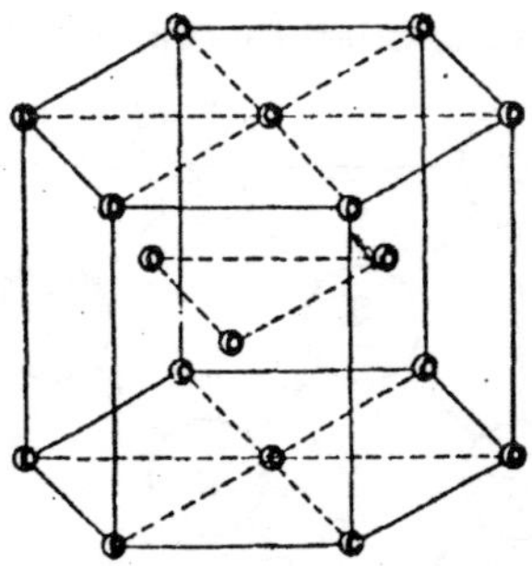

The coordination number in a closed-packed crystal (either fcc or hep) is 12. In both cases each sphere in a particular layer fits into the hollow formed by three spheres in the layer below it; hence each sphere is in contact with three spheres in the layer below it, and with three in the layer above it as well. Adding these six spheres to the six it touches in its own layer, each sphere in a close-packed structure has 12 nearest neighbours,,; that is, the coordination number is 12.

Beryllium, magnesium, cobalt and zinc are among the elements with hcp structures.

In a face-centered cubic (fcc) structure, also called as cubic close-packed, the sequence repeats itself every three layers, instead of every two layers. Copper, lead, gold and argon are examples of close-packed fcc structures.

Diamond Structure

The space lattice of diamond is face-centered cubic (fcc) with a basis of two carbon atoms associated with each lattice point. The figure below shows the positions of atoms in the cubic cell of the diamond structure projected on a cubic face. The fractions denote height above the base in units of a cube edge. The points at 0 and ½ are on the fcc lattice, those at ¼ and ¾ are on a similar lattice displaced along the body diagonal by one-fourth of its length. Thus the diamond lattice is composed of two inter-leaved fcc sublattices, one of which is shifted relative to the other by one-fourth of a body diagonal.

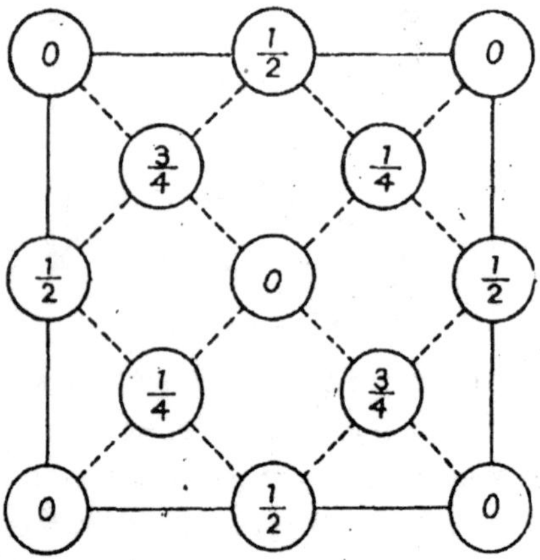

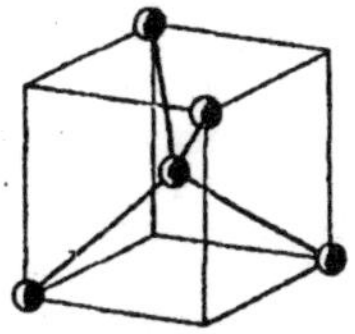

TRA HEDRAL
STRUCTURE

(a)

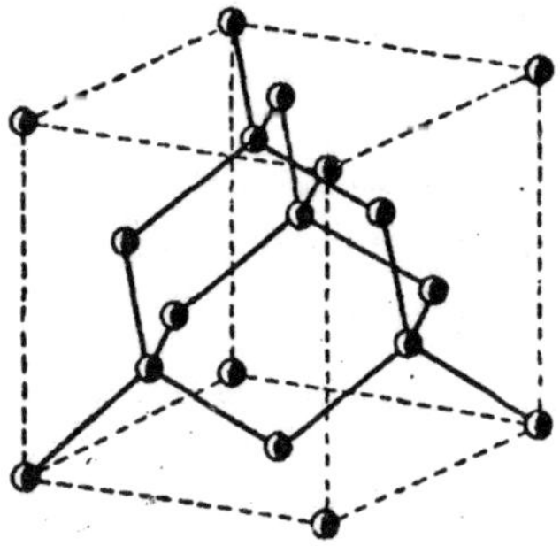

DIAMOND LATTICE

(b)

In a diamond crystal the carbon atoms are linked by directional covalent bonds. Each carbon atom forms covalent bonds with four other carbon atoms that occupy four corners of a cube in a tetrahedral structure (Fig. a). The length of each bond is 1.54 Å and the angle between the bonds is 109.5°. The entire diamond lattice is constructed of such tetrahedral units (Fig. b).

In the diamond lattice, each atom has four nearest neighbours with which it forms covalent bonds. Thus the coordination number of diamond crystal is 4. The number of atoms per unit cell is 8.

Silicon, germanium and gray tin crystallise in the diamond structure.

PROBLEMS

1. ***The lattice constant of NaCl crystal is calculated. The density of NaCl is 2189 kg/m^3 and Avogadro's number N is 6.02 × 10^{26}/kg-molecule.***

Solution : NaCl is a cubic crystal having an fcc lattice. The lattice constant of a cubic lattice is given by

$$a = \left(\frac{nM}{N\rho}\right)^{\frac{1}{3}},$$

where M is molecular weight and ρ is the density of the crystal. n is number of molecules per unit cell.

The molecular weight of NaCl is 58.5 kg/kg-molecule (or 58.5 gm/gm-molecule). Since it belongs to fcc lattice, the number of molecules (in fact, $Na^+ - Cl^-$ ion pairs) per unit cell is 4.

Substituting the values of n, M, N, ρ in the above relation, we have

$$a = \left[\frac{4 \times 58.5 \text{ kg}/(\text{kg-molecule})}{6.02 \times 10^{26}/(\text{kg-molecule}) \times 2189 \text{ kg}/\text{m}^3}\right]^{\frac{1}{3}}$$

$$= (177 \times 10^{-30} \text{ m}^3)^{1/3}$$

$$= 5.61 \times 10^{-10} \text{ m} = 5.61 \text{ Å}.$$

2. ***The lattice constant of KBr from the following data is calculated as follows: density = 2.7 gm/cm*^{3}*, molecular weight = 119, Avogadro's number = 6.02 × 10*^{23}*/gm-molecule. KBr is fcc lattice.***

Solution : The lattice constant for a cubic lattice is given by

$$a = \left(\frac{nM}{N\rho}\right)^{\frac{1}{3}}$$

where n is number of lattice points (atoms or molecules) per unit cell. For fcc lattice, $n = 4$.

$$\therefore\ a = \left[\frac{4 \times 119 \text{ gm}/(\text{gm-molecule})}{6.02 \times 10^{23}/(\text{gm-molecule}) \times 2.7 \text{ gm}/\text{cm}^3}\right]^{\frac{1}{3}}$$

$$= [293 \times 10^{-24} \text{ cm}^3]^{1/8}$$

$$= 6.64 \times 10^{-8} \text{ cm} = 6.64 \text{ Å}.$$

3. ***Copper has a density of 8.96 gm/cm*^{3} *and an atomic weight of 63.5. The distance between two nearest copper atoms***

in the fcc structure is calculated. The Avogadro number is 6.02×10^{23}.

Solution : The lattice constant a for a cubic lattice is given by

$$a = \left(\frac{nM}{N\rho}\right)^{\frac{1}{3}},$$

where n is number of atoms per unit cell, M is atomic weight and ρ is density. For fcc structure, $n = 4$.

$$\therefore \ a = \left[\frac{4 \times 63.5 \text{ gm}/(\text{gm-atom})}{6.02 \times 10^{23}/(\text{gm-atom}) \times 8.96 \text{ gm}/\text{cm}^3}\right]^{\frac{1}{3}}$$

$$= [47.1 \times 10^{-24} \text{ cm}^3]^{1/3}$$

$$= 3.61 \times 10^{-8} \text{ cm} = 3.61 \text{ Å}.$$

The nearest-neighbour distance in fcc lattice is $a/\sqrt{2}$. Therefore, the distance between two nearest copper atoms

$$= \frac{3.61 \text{ Å}}{\sqrt{2}} = 2.55 \text{ Å}.$$

4. ***The density of α-iron is 7870 kg/m^3 and its atomic weight is 55.8. Given that α-iron crystallises in bcc space lattice, its lattice constant is deduced as follows. Avogadro's number N = 6.02×10^{26} per kg-atom.***

Solution : The lattice constant for a cubic crystal of atomic weight M and density ρ is given by

$$a = \left(\frac{nM}{N\rho}\right)^{\frac{1}{3}},$$

where n is the number of atoms per unit cell. For bcc lattice, $n = 2$.

$$\therefore \; a = \left[\frac{2 \times 55.8 \text{ kg/(kg-atom)}}{6.02 \times 10^{26}/(\text{kg-atom}) \times 7870 \text{ kg/cm}^3}\right]^{\frac{1}{3}}$$

$$= [23.5 \times 10^{-30} \text{ m}^3]^{1/3}$$

$$= 2.86 \times 10^{-10} \text{ m} = 2.86 \text{ Å}.$$

5. ***Calculate the lattice constant for CSCI (bcc lattice). Given: density = 8000 kg/m^3, molecular weight t = 168.4 and N=6.02 × 10^{24} per kg-molecule.***

[**Ans.** 4.11 Å]

6. ***The molecular weight of NaCl is 58.5 and its density is 2.17 gm/cm^3. The distance between the adjacent atoms in NaCl crystal is calculated. The Avogadro number is N = 6.02 × 10^{22} per gm-molecule.***

Solution : In the rock-salt (NaCl) crystal, the Na and Cl atoms (strictly ions) occupy alternately the corners of an elementary cube. Let d cm be the distance between adjacent atoms. Then the number of atoms in 1 cm length of an edge of the cube is $1/d$. The total number of atoms in 1 cm^8 (unit volume) is $(1/d)^3$.

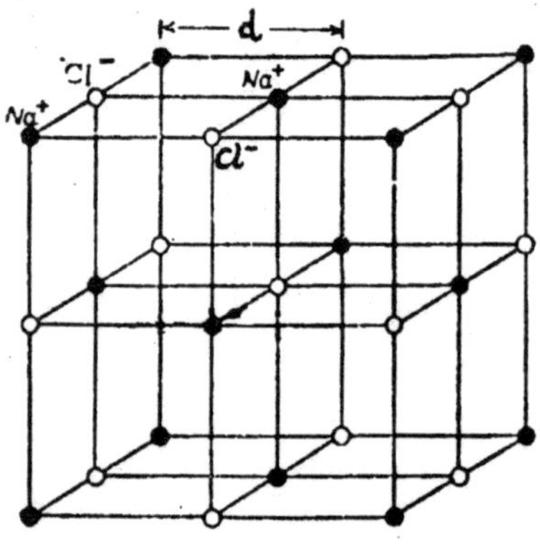

Now, 1 gm-molecule of NaCl contains N molecules and has a mass of M gm, where N is Avogadro's number and M

is the molecular weight of NaCl. Thus N molecules of NaCl have a mass M gm and occupy a volume given by

$$V = \frac{M}{\rho}$$

where ρ is the density of NaCl. Thus the number of molecules per unit volume is N/V. Since there are two atoms in each molecule, the total number of atoms per unit volume is $2N/V$. Hence

$$\left(\frac{1}{d}\right)^3 = \frac{2N}{V} = \frac{2N\rho}{M}$$

or $$d = \left(\frac{M}{2N\rho}\right)^{\frac{1}{3}} \text{ cm.}$$

Substituting the given values :

$$d = \left[\frac{58.5 \text{ gm}/(\text{gm - molecule})}{2 \times (6.02 \times 10^{23}/\text{gm - molecle}) \times 2.17 \text{ gm}/\text{cm}^3}\right]^{\frac{1}{3}}$$

$$= (22.4 \times 10^{-24} \text{ cm}^3)^{1/3}$$

$$= 2.82 \times 10^{-8} \text{ cm} = 2.82 \text{ Å}.$$

7. ***In a crystal, a lattice plane cuts intercepts of a, 2b and 3c along the three axes where a, b, c are primitive vectors of the unit cell. The Miller indices of the given plane is determined as follows.***

Solution : From the law of rational indices, we have

$$a : 2b : 3c = \frac{a}{h} : \frac{b}{k} : \frac{c}{l},$$

where h, k, l are the Miller indices. Thus

$$\frac{1}{h} : \frac{1}{k} : \frac{1}{l} = 1 : 2 : 3$$

$$h : k : l = 1 : \frac{1}{2} : \frac{1}{3}.$$

Converting to smallest whole numbers having the same ratios, we get

$$h : k : l = \frac{6}{6} : \frac{3}{6} : \frac{2}{6} = 6 : 3 : 2$$

Thus $h = 6$, $k = 3$ and $l = 2$, Hence the Miller indices of the plane are 6, 3 and 2; or the plane is (632).

8. ***The Miller indices of a plane which cuts off intercepts in the ratio 1a : 3b : – 2c along the three axes where a b c are primitives is deduced as follows.***

Solution : From the law of rational indices, we have

$$1a : 3b : -2c = \frac{a}{h} : \frac{b}{k} : \frac{c}{l}$$

where h, k, l are the Miller indices. Then

$$\frac{1}{h} : \frac{1}{k} : \frac{1}{l} = 1 : 3 : -2$$

or

$$h : k : l = 1 : \frac{1}{3} : -\frac{1}{2}$$

$$= 6 : 2 : -3.$$

Thus $h = 6$, $k = 2$, $l = -3$. Hence the plane is (623).

9. The Miller indices of a set of parallel planes which make intercepts in the ratio 3a : 4b on the X and Y axes and are parallel to the Z axis. a, b, c are the primitive vectors of the lattice is found as follows.

Solution : The parallel planes are parallel to the Z-axis. This means that their intercepts on the Z-axis are infinite.

From the law of rational indices, we have

$$3a : 4b : \infty\, c = \frac{a}{h} : \frac{b}{k} : \frac{c}{l}$$

or $$\frac{1}{h}:\frac{1}{k}:\frac{1}{l} = 3 : 4 : \infty$$

$$h : k : l = \frac{1}{3}:\frac{1}{4}:\frac{1}{\infty} = 4 : 3 : 0.$$

The Miller indices are 4, 3, 0.

10. ***In a crystal whose primitives are 1.2 Å, 1.8 Å and 2.0 Å, a plane (231) cuts an intercept 1.2 Å on the X-axis. The corresponding intercepts on the Y- and Z-axis are found as follows.***

Solution : Let p, q and r be the intercepts on the X–, Y–, and Z-axis respectively. Then, from the law of rational indices, we write

$$p : q : r = \frac{a}{h}:\frac{b}{k}:\frac{c}{l}$$

where a, b, c are the primitives and h, k, l are the Miller indices. Here a = 1.2 Å, b =1.8 Å and c = 2.0 Å, and h =2, k = 3 and l = 1. Thus

$$p : q : r = \frac{1.2}{2}:\frac{1.8}{3}:\frac{2.0}{1}$$

$$= 0.6 : 0.6 : 2.0$$

But p = 1.2 Å.

$$\therefore \quad 1.2 : q = 0.6 : 0.6$$

$$\therefore \quad q = 1.2 \text{ Å}$$

Similarly $1.2 : r = 0.6 : 2.0$

$$\therefore \quad r = \frac{1.2 \times 2.0}{0.6} = 4.0 \text{ Å}$$

Thus the intercepts on Y and Z axes are 1.2 Å and 4.0 Å respectively.

11. ***The ratio of intercepts on the three axes by (132) planes in a simple cubic lattice is found as follows.***

Solution : Let p, q, r be the intercepts on the X–, Y– and Z–. axis respectively. Then,

$$p : q : r = \frac{a}{h} : \frac{b}{k} : \frac{c}{l}$$

where a, b, c are the primitives and h, k, l are the Miller indices. Here $h = 1$, $k = -3$ and $l = 2$; also $a = b = c$ for the "cubic" lattice.

$$\therefore\ p : q : r = \frac{a}{1} : \frac{a}{-3} : \frac{a}{2} = 6 : -2 : 3$$

12. *The lattice constant for a cubic lattice is a. The spacings between (011), (101) and (112; planes are deduced as follows.*

Solution : For a 'cubic' lattice, the interplanar spacing d is given by

$$d = \frac{a}{\sqrt{\left(h^2 + k^2 + l^2\right)}},$$

where h, k, l are the Miller indices.

For (011) plane, $h = 0$, $k = 1$, $l = 1$.

$$\therefore \qquad d_{011} = \frac{a}{\sqrt{\left[(0)^2 + (1)^2 + (1)^2\right]}} = \frac{a}{\sqrt{2}}$$

Similarly, for (101) plane,

$$d_{101} = \frac{a}{\sqrt{\left[(1)^2 + (0)^2 + (1)^2\right]}} = \frac{a}{\sqrt{2}}$$

and for (112) plane,

$$d_{112} = \frac{a}{\sqrt{\left[(1)^2 + (1)^2 + (2)^2\right]}} = \frac{a}{\sqrt{6}}$$

13. *The interplanar spacing for a (321) plane in a simple cubic lattice whose lattice constant is 4.2 x 10^{-8} cm is calculated as follows.*

Solution : In a simple cubic lattice the interplanar spacing d is given by

$$d = \frac{a}{\left(h^2 + k^2 + l^2\right)^{1/2}}$$

where h, k, l are the Miller indices. For a (321) plane, we have $h = 3$, $k = 2$ and $l = 1$. Also, here $a = 4.2 \times 10^{-8}$ cm.

$$\therefore \quad d = \frac{4.2 \times 10^{-8}\,\text{cm}}{\left(3^2 + 2^2 + 1^2\right)^{1/2}}$$

$$= \frac{4.2 \times 10^{-8}\,\text{cm}}{\sqrt{14}} = \frac{4.2 \times 10^{-8}\,\text{cm}}{3.74}$$

$$= 1.12 \times 10^{-8}\ \text{cm} = 1.12\ \text{Å}.$$

14. ***In a tetragonal lattice a = b = 2.5 Å, c = 1.8 Å. The lattice spacing between (111) planes is deduced as follows.***

Solution : The general expression for the interplanar spacing d is

$$d_{hkl} = \frac{1}{\sqrt{\left(\frac{h^2}{a^2} + \frac{k^2}{b^2} + \frac{l^2}{c^2}\right)}}$$

where h, k, l are the Miller indices and a, b, c are the primitives.

Here $h = 1$, $k = 1$, $l = 1$, $a = b = 2.5$ Å, $c = 1.8$ Å.

$$\therefore \quad d_{111} = \left[\frac{1}{\left(2.5\ \text{Å}\right)^2} + \frac{1}{\left(2.5\ \text{Å}\right)^2} + \frac{1}{\left(1.8\ \text{Å}\right)^2}\right]^{-\frac{1}{2}}$$

$$= \left[\frac{1}{6.25} + \frac{1}{6.25} + \frac{1}{3.24}\right]^{-\frac{1}{2}} \text{Å}$$

$$= [0.16 + 0.16 + 0.31]^{-1/2} \text{ Å}$$

$$= (0.63)^{-1/2} = \left(\frac{100}{63}\right)^{1/2} = 1.26 \text{ Å}$$

15. ***For a simple cubic crystal find (i) the ratio of intercepts on the three axes by (123) plane, (ii) the ratio of the spacings of (110) and (111) planes and (iii) the ratio of the nearest neighbour distance to the next nearest neighbour distance.***

Solution : (*i*) Let p, q, r be the intercepts on the X-, Y- and Z- axis respectively. Then, from the law of rational indices, we have

$$p : q : r = \frac{a}{h} : \frac{b}{k} : \frac{c}{l},$$

where a, b, c are primitives and h, k, l are Miller indices.

Here $h = 1, k = 2, l = 3$; also $a = b = c$ for a "cubic" lattice.

$$\therefore \qquad p : q : r = \frac{a}{1} : \frac{a}{2} : \frac{a}{3} = 6 : 3 : 2$$

(ii) The spacing between (hkl) planes in a cubic lattice is given by

$$d_{hkl} = \frac{a}{\left(h^2 + k^2 + l^2\right)^{1/2}}.$$

Thus
$$d_{110} = \frac{a}{\left(1^2 + 1^2 + 0^2\right)^{1/2}} = \frac{a}{\sqrt{2}}$$

and
$$d_{111} = \frac{a}{\left(1^2 + 1^2 + 1^2\right)^{1/2}} = \frac{a}{\sqrt{3}}$$

$$\frac{d_{110}}{d_{111}} = \frac{a/\sqrt{2}}{a/\sqrt{3}} = \sqrt{3} : \sqrt{2}.$$

(iii) In a "simple" cubic lattice the distance between nearest neighbours is a, and that between next nearest neighbours is $\sqrt{2}\,a$ (see Fig.). Their ratio

$$= a/\sqrt{2}a = 1/\sqrt{2}.$$

16. *The densities of lattice points in (111) and (110) planes in a simple cubic lattice is compared.*

Solution : For a simple cubic (primitive) lattice, the (surface) density of lattice points is given by

$$\rho = \frac{d}{abc} = \frac{d}{a^3}, \qquad [\because a = b = c]$$

where d is interplanar spacing. Thus

$$\frac{\rho_{111}}{\rho_{110}} = \frac{d_{111}}{d_{110}}$$

For a simple cubic lattice, $d_{111} = a/\sqrt{3}$ and $d_{110} = a/\sqrt{2}$.

$$\therefore \qquad \frac{\rho_{111}}{\rho_{110}} = \frac{a/\sqrt{3}}{a/\sqrt{2}} = \frac{\sqrt{3}}{\sqrt{2}}$$

or $\qquad P_{111} : P_{110} = \sqrt{2} : \sqrt{3}$

17. *The density of (100) plane in a simple cubic (P) lattice; given a = 2.5 Å.*

Solution : For a cubic P lattice, we have

$$\rho = \frac{d}{a^3}.$$

Now $d_{100} = a$.

$$\therefore \qquad \rho = \frac{a}{a^3} = \frac{1}{a^2}$$

$$= \frac{1}{(2.5 \times 10^{-10}\,\text{m})^2}$$

$$= 1.6 \times 10^{19} \text{ lattice point/m}^2.$$

10

Phase Space

In classical mechanics, the dynamical state of a particle at a particular instant is completely specified if its three position coordinates x, y, z and three velocity, or preferably momentum components p_x, p_y, p_z, at that instant are known. This conception is generalised by imagining a six-dimensional space in which a point has six coordinates (x, y, z, p_n, p_y, p_z). Such a space is called 'phase space'. The instantaneous state (position and momentum) of a particle is represented by a point in the phase space.

For a system containing a large number of particles, there would be a large number of representative points in the phase space corresponding to the instantaneous distribution of particles. Thus the state of a system of particles corresponds to a certain distribution of points in phase space.

Let us divide the phase space into tiny six-dimensional cells whose sides are $dx, dy, dz, dp_x, dp_y, dp_z$. The volume of each cell is

$$d\tau = dx\, dy\, dz\, dp_x\, dp_v\, dp_z.$$

$d\tau$ is termed as 'an element of volume in the phase space'. $dx\, dy\, dz$ is an element of volume in coordinate space and $dp_x\, dp_y\, dp_z$ is an element of volume in momentum space.

Now, according to the uncertainty principle, we have

$$dx\, dp_x \geq \frac{h}{4\pi}\ ;\ dy\, dp_y \geq \frac{h}{4\pi}\ ;\ dz\, dp_z \geq \frac{h}{4\pi}$$

so that $$d\tau \geq \left(\frac{h}{4\pi}\right)^3.$$

A more detailed analysis shows that

$$d\tau = h^3.$$

Thus, due to uncertainty principle, a "point" in phase space is actually a cell of volume h^3. A point (x, y, z, p_x, p_y, P_z) in the phase space representing the position and momentum of a particle specifies that the particle lies somewhere in a cell of volume A^8 centered at the point.

The notion of phase space is important in describing the behaviour of a system of particles. The equilibrium state of the system corresponds to the most probable distribution of particles in the phase space.

Methods of Statistical Mechanics

Every solid, liquid or gas is a system (or an assembly) consisting of" an enormous number of microscopic particles. Likewise, radiation is an assembly of photons. Obviously, the actual motions or interactions of individual particles cannot be investigated. However, the macroscopic properties of such systems can be explained in terms of the most probable behaviour of the individuals. By employing statistical methods we can determine how the individuals of a given assembly are distributed among different possible states and what is their most probable behaviour.

Let us consider statistically how a fixed amount of energy is distributed among the various individuals of an assembly of *identical* particles. There are three kinds of identical particles:

(i) Identical particles of any spin which are so much separated in the assembly that they *can be distinguished* from one another. The molecules of a gas are particles of this kind. The Maxwell-Boltzmann distribution holds for these particles.

(ii) Identical particles of 0 or integral spin which *cannot be distinguished* from one another. These are called Bose particles (or Bosons) and do not obey Pauli's exclusion principle. Photons, phonons and α-particles are of this kind. The Bose Einstein distribution holds for them.

(iii) Identical particles of odd half-integral spin which *can not be distinguished* from one another. These are called Fermi particles (or Fermions) and do obey Pauli's exclusion principle. Electrons, protons and neutrons are particles of this kind. The Fermi-Dirac distribution holds for them.

The energy distribution law for particles of kind (*i*) can be derived by methods of classical statistics as well as of quantum statistics, but that for particles of kinds (*ii*) and (*iii*) can be derived by methods of quantum statistics only.

Maxwell-Boltzmann (Classical) Distribution : Let us consider a system composed of a very large number N of *distinguishable* identical particles. Suppose the energies of the particles are limited to the values $\varepsilon_1, \varepsilon_2, \ldots\ldots, \varepsilon_r$ which represent either discrete energy states or average energies within a sequence of energy intervals. Let us divide the whole volume of the phase space into r cells and distribute the N molecules among these cells. Let us consider a distribution of the molecules with respect to their energy such that n_1, molecules (each having energy ε_1) are in cell 1, n_g molecules (each having energy ε_2) are in cell 2, and so on. The probability W for the distribution is given by

$$W = \frac{N!}{n_1 ! n_2 ! \ldots\ldots . n_r !} (g_1)^{n_1} (g_2)^{n_2} \ldots \ldots (g_r)^{n_r}$$

$$= \frac{N!}{\Pi n_i !} \Pi (g_t)^{n_i,}$$ [just as Σ stands for sum Π stands for product]

where g_i is the a *priori probability* for a particle to have the energy ε_1.

There are two limitations. The total number of particles is fixed at N. This means that

$$\sum_{i-1}^{r} nt = n_1 + n_2 + \ldots \ldots + nr = N.$$

At a constant temperature, the total energy content is fixed at E (say). Thus

$$\sum_{i-1}^{r} nt\ \varepsilon i = n_1 \varepsilon_1 + n_2 \varepsilon_2 + \ldots \ldots + n_r \varepsilon_r = E.$$

The next step is to determine the most probable distribution of the particles which would correspond to the state of thermodynamic equilibrium. This distribution would yield maximum value of W. Thus, for the equilibrium state, we have

$$\delta W = 0$$

subject to the limitations

$$\Sigma \delta n_i = 0$$

and $$\Sigma \varepsilon_i \delta n_i = 0$$

This leads to the result

$$n_i = \frac{gi}{e^{\alpha} e^{\varepsilon i / kT}},$$

where e^{α} is a constant, k is Boltzmann's constant and T is Kelvin temperature. This result is known as Maxwell-Boltzmann distribution law.

Bose-Einstein (Quantum) Distribution : The Maxwell-Boltzmann distribution governs identical particles which can be distinguished from one another in some way (as molecules in a gas) and as such they can be given names or numbers. On the other hand, Bose-Einstein distribution governs identical particles which can never be distinguished from one another (as radiation photons in an enclosure). As such there is no way to number them or give them names.

The basic assumption of the B-E distribution is that any number of particles can be in any quantum state and that all quantum states are equally probable. Suppose each quantum state corresponds to an elementary cell in the phase space. The number of different ways in which n_i *indistinguishable* particles can be distributed in g_i phase-cells is

$$\frac{(n_i + gi - 1)^!}{ni!(gi-1)^!}$$

When the number of cells is sufficiently large, this is same as

$$\frac{(n_i + gi)^!}{ni!\,gi^!}$$

The probability W of the entire distribution of N particles is the *product of* the numbers of different arrangements of particles among the states having each energy. Thus

$$W = \Pi \frac{(n_i + gi)^!}{ni!\,gi^!}$$

For the equilibrium (most probable) condition, we must have

$$\delta W = 0$$

subject to the limitations

$$\Sigma\, \delta n_i = 0$$

and $\quad \sum \varepsilon i \, \delta n_i = 0.$

This leads to the result

$$ni = \frac{gi}{e^{\alpha} e^{\epsilon i/kT} - 1}$$

where e^{α} is a constant, k is Boltzmann's constant and T is Kelvin temperature. This is Bose-Einstein distribution law.

Fermi-Dirac (Quantum) Distribution : Fermi-Dirac distribution applies to *indistinguishable* particles , like electrons, which are governed by Pauli's exclusion principle. The general method of deriving the Fermi-Dirac distribution law is similar to that for the Bose-Einstein statistics except that now each phase-cell (that is, each quantum state) cannot be occupied by more than one particle. This leads to the following distribution function :

$$ni = \frac{gi}{e^{\alpha} e^{\epsilon i/kT} + 1}.$$

Again, n_i is the number of particles whose energy is ε_i, and g_i is the number of states that have the same energy εi.

Comparison of the Three Distribution Laws : For this purpose, let us define a quantity

$$f(\varepsilon_i) = \frac{ni}{gl}.$$

$f(\varepsilon_i)$ is called the 'occupation index' of a state of energy ε_i. It represents the average number of particles in each of the states of that energy. Thus

$$f_{MB}\,(\varepsilon i) = \frac{1}{e^{\alpha} e^{\epsilon i/kT}}$$

$$f_{ME}\,(\varepsilon i) = \frac{1}{e^{\alpha} e^{\epsilon i/kT} - 1}$$

and $$f_{FD}\,(\varepsilon i) = \frac{1}{e^{\alpha} e^{\epsilon i/kT} + 1}$$

In the following diagrams we plot each occupation index $f(\varepsilon_i)$ against energy ε_i for two different values of T and α.

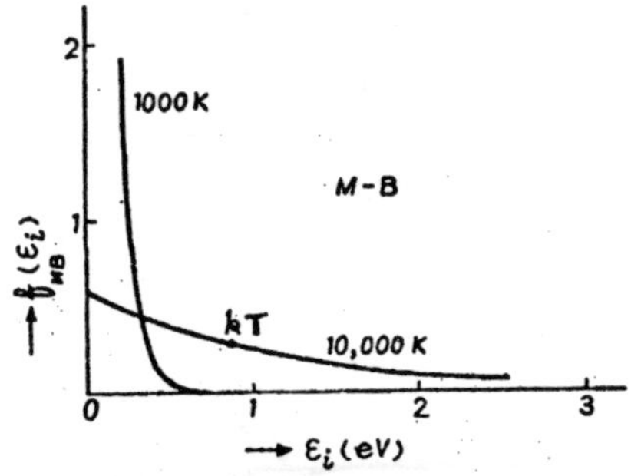

In the M-B distribution (Fig.) the occupation index $f(\varepsilon_i)$ falls purely *(exponentially* with increase in energy ε_i. In fact , it falls by a factor of $\frac{1}{e}$ for each increase of kT in the energy $\in_i$.

The B-E occupation index against energy is plotted in the figure below for temperatures of 1000 K and 10,000 K, in each case for $\alpha = 0$ (corresponding to a "gas" of photons). For $\in_i >> kT$, the B-E distribution approaches the exponential form characteristic of the M-B distribution. In this region the average number of particles per quantum state is much less than one. However, for $\in_i << kT$, the — 1 term in the denominator causes the occupation index much greater compared to that of the M-B distribution. This means that for energies small compared to kT, the number of particles per quantum state is greater for the B-E distribution than for the M-B distribution.

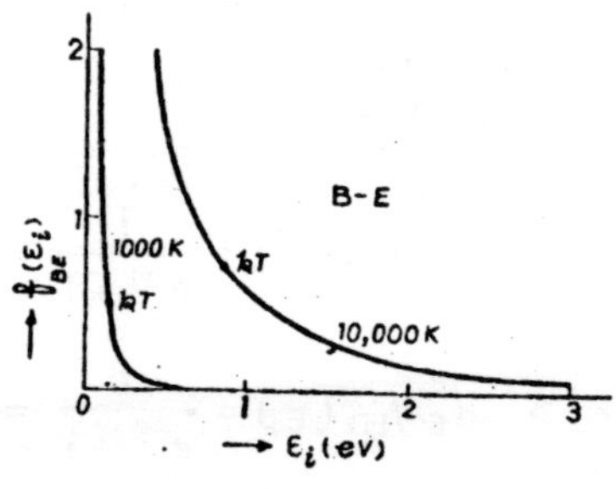

The F-D distribution is plotted in Fig. below for three different values of T and α. In this distribution the occupation index never goes above 1. This signifies that we cannot have more than one particle per quantum state as required by Pauli's exclusion principle which applies in this case. Further, in this distribution, the parameter α is strongly dependent on temperature T, and we write

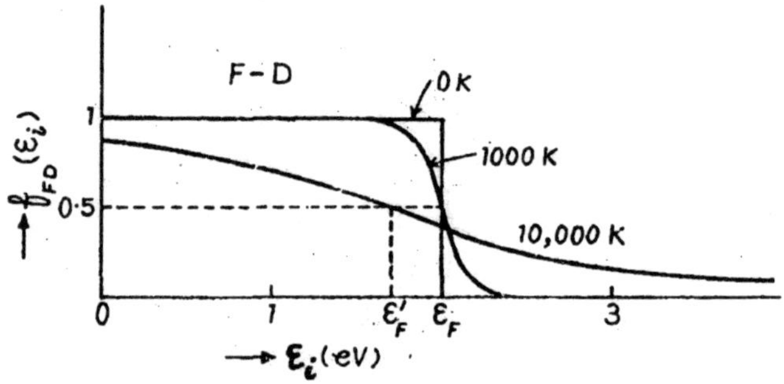

$$\alpha = -\frac{\in F}{kT}$$

so that the F-D occupation index becomes

$$f_{FD}(\in i) = \frac{1}{e^{(\in i - \in F)/kT} + 1},$$

where $\in_F$ is called the 'Fermi-energy'.

Let us consider the situation at the absolute zero of temperature. At $T = 0$, $(\in_i - \in_F)/kT = -\infty$ (for $\in_i < \in_F$) and $+\infty$ (for $\in_i > \in_F$), so that

$$\text{for } \in_i < \in_F, f_{FD}(\in_i) = \frac{1}{e^{-\infty} + 1} = 1, \qquad (\because e^{-\infty} = 0)$$

$$\text{and for } \in_i > \in_F, f_{FD}(\in_i) = \frac{1}{e^{+\infty} + 1} = 0. \qquad (\because e^{+\infty} = \infty)$$

Thus, at $T = 0$, all energy states from $\in_i = 0$ to $\in_i = e_F$ are occupied because $f_{FD}(\in_i) = 1$; while all above $\in_F$ are vacant.

As the temperature rises, some of the states just below $\in_F$ become vacant while some just above $\in_F$ are occupied. The higher the temperature, the more is the spreading in $f_{FD}(\in_i)$.

At $\in_i = \in_F$, we have

$f_{FD}(\in_i)\ \dfrac{1}{e^0 + 1} = ½$, at all temperatures, that is, the average number of particles per quantum state is exactly ½ . In other words, *the probability of finding an electron with energy equal to the Fermi energy in a metal is ½ at any temperature.*

11

Monatomic Linear Crystal

Vibrations of Monatomic Linear Lattices

A monatomic linear crystal lattice is an array of identical, equidistant atoms (or ions), each of mass M (say). When an elastic wave propagates through the crystal, the atoms are set in vibration either parallel or perpendicular to the direction of the wave vector. We shall consider longitudinal elastic vibrations of the atoms. We assume that interaction exists only between nearest neighbours and that Hooke's law is obeyed.

Below given figure shows a linear chain of identical atoms. Let it be infinitely long. The open circles represent the equilibrium positions and the black dots the displaced atoms. Let a be the equilibrium interatomic spacing, and x_0, x_1, x_2, .. x_{n-1} . x_n, x_{n+1}, ... the displacements from the equilibrium positions. According to Hooke's law, the force on an atom due to the displacement of its (nearest) neighbour is proportional to the difference of their displacements. So the net force on the nth atom is

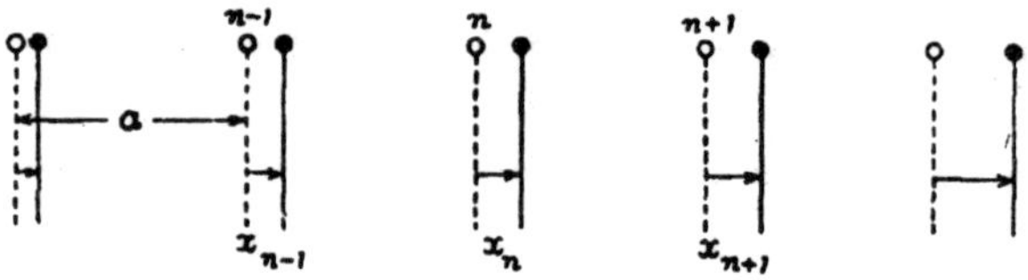

$$F_n = f(x_{n+1} - x_n) - f(x_n - x_{n-1})$$
$$= f(x_{n+1} + x_{n-1} - 2x_n)$$

where f is the force constant for the nearest-neighbour interaction. The equation of motion of the nth atom is then

$$M x_n = f(x_{n+1} + x_{n-1} - 2x_n) \qquad ...(i)$$

Let a solution to this equation be of the form of a travelling waver

$$x_n = e^{-iw(t - na/v)},$$

where na is the equilibrium position (x-coordinate) of the nth particle relative to the origin, and v the velocity of wave propagation.

Let us introduce the wave vector K, where

$$K = \frac{2\pi}{\lambda} = \frac{\omega}{v}. \quad [\therefore \omega = 2\pi v \text{ and } v = v\lambda]$$

Then we have

$$x_n = e^{-i(wt - Kna)}. \qquad ... (ii)$$

Accordingly,

$$x_{n+1} = e^{-i\{wt - K(n+1)a\}} = e^{iKa} x_n$$

and

$$x_{n-1} = e^{-i\{wt - K(n-1)a\}} = e^{-iKa} x_n.$$

Differentiating eq. (ii) twice with respect to t, we get

$$\ddot{x}_n = \frac{d^2 x_n}{dt^2} = (-iw)2\, e^{-i(wt - Kna)} = -w^2 x_n.$$

Substituting for $\ddot{x}_n$, x_{n+1}, x_{n-1} and x_n in eq. (i), and dividing by x_n, we get

$$-M\omega^2 = f(e^{iKa} + e^{-iKa} - 2)$$
$$= f(e^{iKa/2} - e^{-iKa/2})^2.$$

Now, $\sin \dfrac{Ka}{2} = \dfrac{e^{iKa/2} - e^{-iKa/2}}{2i},$

and so, $\sin^2 \dfrac{Ka}{2} = -\dfrac{1}{4}(e^{iKa/2} - e^{-iKa/2})^2.$

$$\therefore \qquad -M\omega^2 = -4f \sin^2 \frac{Ka}{2}$$

or $$\omega^2 = \frac{4f}{M}\sin^2 \frac{Ka}{2}$$

or $$\omega = 2\sqrt{\frac{f}{M}}\left|\sin \frac{Ka}{2}\right|. \qquad \text{... } (iii)$$

We take the sign of the square-root so that the angular frequency ω is always positive.

We have thus obtained an expression for the frequency ω of the elastic waves in a monatomic linear lattice in terms of K (that is, in terms of A). The ω-K relationship is shown in Fig. below. It shows that frequency is a periodic function of K, with period $2n/a$. The dashed lines correspond to a continuous string for which $\omega \propto K$.

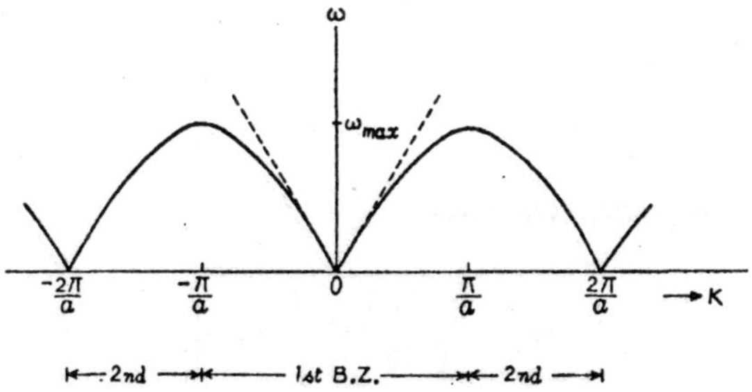

Results : We draw the following results from the above discussion:

(1) For values of K much smaller than interatomic spacing a, that is, for wavelengths large compared with a, we have

$$\sin \frac{Ka}{2} \simeq \frac{Ka}{2} \qquad [\therefore\ Ka < 1]$$

$$\therefore \qquad \omega = 2\sqrt{\frac{f}{M}}\left(\frac{Ka}{2}\right) = \alpha \sqrt{\frac{f}{M}}(K).$$

The result that the frequency is directly proportional to the wave vector is a statement that the velocity of propagation of waves is independent of wavelength. That is, dispersion effects are negligible. This is what happens in a continuous string. Thus, *for long wavelengths,* the periodicity of the lattice is of little importance.

As K increases (wavelength decreases), however, the dispersion effects become important and ω is no longer directly proportional to K.

(2) Another important result is obtained by comparing the solution (it) with another in which K is replaced by K_m where

$$Km = K + \frac{2\pi m}{a} \text{ with } m = \pm 1, \pm 2, \ldots \ldots$$

We have for the displacement of the nth atom

$$x_n = e^{-\iota(\omega t - K_m na)}$$

$$= e^{-\{\omega t - (K + 2\pi m/a)\, na\}}$$

$$= e^{-i(\omega t - Kna)}, \qquad [\therefore\ e^{2\pi im} = 1]$$

and from eq. (*iii*), we have

$$\omega = \sqrt{\frac{4f}{M}\left|\sin \frac{1}{2} K_m a\right|}$$

$$= \sqrt{\frac{4f}{M}\left|\sin \frac{1}{2} a\left(K + \frac{2\pi m}{a}\right)\right|}$$

$$= \sqrt{\frac{4f}{M}\left|\sin\frac{1}{2}Ka\right|} .$$

Thus the solutions (*ii*) and the frequencies (*iii*) corresponding to K and K_m are identical. It means that the state of vibration of the lattice corresponding to a wave vector K is the same as that for any of the wave vectors $K + 2\pi m/a$. Therefore, to find a unique relation between the state of vibration and the wave vector, the wave vector K must be restricted to a range of values $2\pi/a$. We specify the range of independent values of K by

$$-\frac{\pi}{a} \leq K \leq \frac{\pi}{a} .$$

We have taken both positive and negative values of K because waves can propagate to the right or to the left. This range of K values is referred to as the "first Brillouin zone" of the linear lattice. The extreme values of K in this zone are

$$K_{max} = \pm \frac{\pi}{a},$$

where K_{max} may be of the order of 108 cm^{-1}. (In an elastic continuum $a \to 0$ and $K_{max} \to \pm \infty$).

At the boundaries $K_{max} = \pm \frac{\pi}{a}$ of the Brillouin zone the solution $x_n = e^{-i(\omega t - Kna)}$ becomes

$$x_n = e^{-i\omega t} e^{\pm in\pi} = e^{-i\omega t} (-1)n$$

which represents a "standing" wave (rather than a travelling wave) in which alternate atoms vibrate in opposite phases. This situation is equivalent to Bragg reflection of X-rays. When the Bragg condition is satisfied, a wave cannot propagate in a lattice, but through successive reflections back and forth , a standing wave is set up. Here the critical value $K_{max} = \pm \frac{\pi}{a}$ satisfies the Bragg condition $2d \sin\theta = n\lambda$. Here we have $\theta = \frac{\pi}{2}$ (linear lattice), $d = a, n = 1$, so that

$$\lambda = \lambda_{\min} = 2a$$

which corresponds to $K_{max} = \pm \pi/a$.

3. The frequency corresponding to zone boundaries ($K = K_{max} = \pm \pi/a$) is, according to eq. (*iii*),

$$\omega_{max} = 2\sqrt{\frac{f}{M}}$$

or $$V_{max} = \frac{1}{\pi}\sqrt{\frac{f}{M}}. \qquad [\because \omega = 2\pi v]$$

Thus there exists a maximum frequency v_{max} which can be propagated through the chain. Hence a monatomic linear lattice acts as a low-pass filter which transmits only in the frequency range between zero and v_{max}.

$$v_{max} = \frac{v}{\lambda_{\min}} = \frac{v}{2a}.$$

Now a $\simeq 10^{-8}$ cm and the velocity of sound in solids is $\sim 10^5 - 10^6$ cm/sec, so that

$$v_{max} \simeq 10^{13}\ \text{sec}^{-1}.$$

Hence the cut-off frequency lies in the infrared region.

In contrast with this, the continuous string has no frequency limit.

Vibrational Modes of a Finite Monatomic Linear Lattice : Let us consider an array of $(N + 1)$ identical atoms , numbered from 0 to N. Suppose that the two end atoms are fixed, so that $(N - 1)$ atoms are mobile. Let a be the equilibrium interatomic spacing , and $x_0, x_1, x_2, \ldots\ldots x_N$ the instantaneous displacements from the equilibrium positions. Assuming Hooke's law to be obeyed and considering only nearest-neighbour interaction, the equation of motion of the nth atom is

$$M\ddot{x}_n = f(x_{n+1} - x_n) - f(x_n - x_{n-1})$$

$$= f(x_{n+1} + x_{n-1} - 2x_n) \qquad \ldots (i)$$

where M is the mass of an atom and f is the force constant for the nearest-neighbour interaction.

The general solution of eq. (*i*) can be written as a sum of two travelling waves, one propagating to the right and the other to the left :

$$x_n = [A_1 e^{i(Kna + \beta_1)} + A_2 e^{i(-Kna + \beta_2)}] e^{-i\omega t}, \quad \text{... } (ii)$$

where A_1 and A_2 are amplitudes, β_1 and β_2 are phase angles and K is wave vector ($K = 2\pi/\lambda$). The boundary conditions are

$$x_0 = 0$$

$$x_N = 0 \text{ for all } t.$$

The first condition, when substituted in (it), requires

$$A_1 = -A_2$$

and $\quad \beta_1 = \beta_2.$

Since the phase angles are equal, we can take $B_1 = B_2 = 0$. The remaining solution in (*ii*) is

$$x_n = A_1 (e^{iKna} - e^{-ikna}) e^{-i\omega t}.$$

Taking the real part only, we have

$$x_n = 2A_1 \sin Kna \sin \omega t. \quad \text{... } (iii)$$

This represents a "standing" wave. It leads to the same dispersion relation (ω-K relationship)

$$\omega^2 = \frac{4f}{M}\sin^2\frac{Ka}{2}$$

as the travelling wave solution, with the difference that K is limited to positive values ranging from 0 to π/a.

The second boundary condition imposed on (*iii*) gives

$$0 = 2A_1 \sin KNa \sin \omega t$$

or $\quad \sin KNa = 0$

or $\quad K = \frac{\pi}{Na} j, \quad \text{... } (iv)$

where j is an integer. However, $j = 0$ must be excluded because it corresponds to $K = 0$ which means that all atoms are at rest.

The maximum value of K is $\frac{\pi}{a}$ which gives $j_{max} = N$. This value must also be excluded because it yields in (*iii*) $x_n = 0$ for all t, that is, again all atoms are at rest. We therefore conclude that

$$j = 1, 2, 3, \ldots\ldots\ldots, (N - 1). \qquad \ldots (v)$$

Thus the boundary conditions select a discrete set of K values given by (*iv*) and (*v*). We see that, *there are just as many modes of vibration (K-values) as there are mobile atoms.*

To each value of K there corresponds a frequency ω_K. Hence the frequency spectrum consists of $(N - 1)$ discrete lines.

Vibrational Modes of a Diatomic Linear Lattice: For crystals with more than one atom per primitive cell, the vibrational spectrum shows new features. For each polarisation mode in a given direction of propagation, the dispersion relation $\omega - K$ gives rise to two branches, known as 'acoustical branch and optical branch'.

Let us consider a linear diatomic lattice, with atoms of two kinds arranged alternately (Fig.). Let the atoms of mass M_1 be even-numbered, and those of mass M_2, odd-numbered, and Let a be the distance between nearest neighbours, so that $2a$ is the repeat distance.

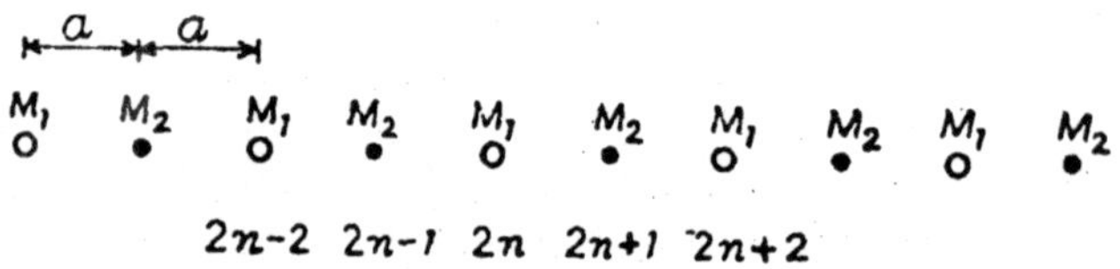

We assume that each atom interacts only with its nearest neighbour and that the force constants are identical between all nearest-neighbour pairs. The equations of motion are

$$M_1 \ddot{x}_{2n} = f(x_{2n+1} + x_{2n-1} - 2x_{2n})$$

and $M_2 \ddot{x}_{2n+1} = f(x_{2n+2} + x_{2n} - 2x_{2n+1}).$ ... (*i*)

where f is the force constant and x_{2n} is the displacement of the 2*n*th particle from the equilibrium position.

Let the solutions to these equations be of the form of a travelling wave :

$$x_{2n} = A\, e^{-i\,(\omega t - 2nKa)}$$

and $$x_{2n+1} = B\, e^{-i\,\{\omega t - (2n+1)Ka\}} \quad \text{... } (ii)$$

where K is the wave vector of a particular mode of vibration, and A and B are the amplitudes corresponding to the atoms M_1 and M_1 respectively. Substituting (*ii*) in (*i*), we get

$$(M_1\,\omega^2 - 2f)A + 2\,Bf\cos Ka = 0\;)$$

and $$(M_2\,\omega^2 - 2f)\,B + 2Af\cos Ka = 0.... \; (iii)$$

These equations have non-vanishing solutions for A and B only if the determinant of the coefficients of A and B vanishes, that is,

$$\begin{vmatrix} (M_1\,\omega^2 - 2f) & 2f\cos Ka \\ 2f\cos Ka & (M_2\,\omega^2 - 2f) \end{vmatrix} = 0$$

or $$(M_1\omega^2 - 2f)\,(M_2\omega^2 - 2f) - 4f^2\cos^2 Ka = 0$$

or $$M_1 M_2\,\omega^4 - 2f\omega^2)\,(M_1 + M_2) + 4f^2\,(1 - \cos^2 Ka = 0$$

giving finally

$$\omega^2 = f\left(\frac{1}{M_1} + \frac{1}{M_2}\right) \pm f\sqrt{\left[\left(\frac{1}{M_1} + \frac{1}{2}\right)^2 - \frac{4\sin^2 Ka}{M_1 M_2}\right]} \quad \text{... } (iv)$$

This is the dispersion relation for the linear diatomic lattice. The form of the relation shows that, as in the monatomic case, the frequency ω is a periodic function of the wave vector K.

Since ω should be positive, each value of ω^2 leads to a single value for *at*. Thus, in contrast to the monatomic lattice, there are now two angular frequencies ω_+ and ω_- corresponding to a single value of K. In a $\omega - K$ plot (Fig.) this leads to two "branches", the one corresponding to ω_+ is called the 'optical' branch and that corresponding to ω_- is called the 'acoustical' branch.

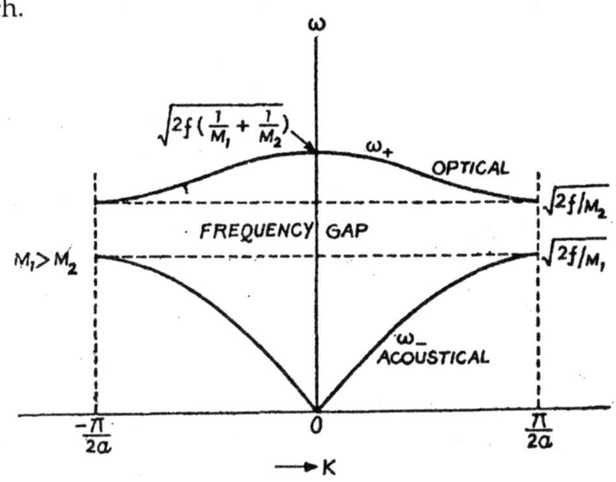

Let us discuss the two branches. For $K = 0$, two roots of (*iv*) are

$$\omega + = \sqrt{\left[2f\left(\frac{1}{M_1}+\frac{1}{M_2}\right)\right]} \qquad \ldots (v)$$

and $\qquad \omega - = 0.$

The range of the first Brillouin zone is

$$\frac{\pi}{2a} \leq K \leq \frac{\pi}{2a},$$

where *la* is the repeat distance of the lattice. At $K = K_{max} = \pm \frac{\pi}{2a}$, the angular frequencies are

$$\omega + = \sqrt{\left(\frac{2f}{M_1}\right)}, \ \omega - = \sqrt{\left(\frac{2f}{M_2}\right)}.$$

The larger the mass ratio M_1/M_2, the wider the 'frequency gap' between the two branches. The existence of a frequency (or energy) gap is a characteristic feature of elastic waves in diatomic lattices.

Let us now investigate the physical difference between the optical and acoustical branches. The particle displacements for the optical and acoustical modes of vibration of (transverse) waves in diatomic linear lattice are shown in Fig. below.

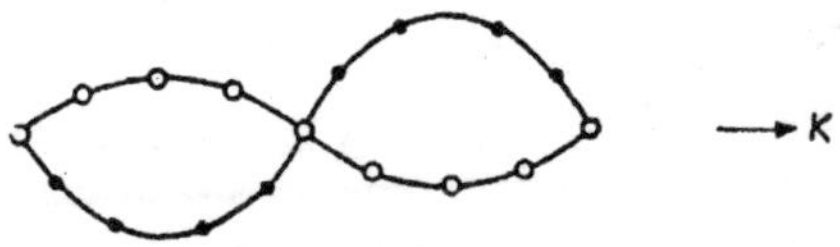

OPTICAL MODE

ACOUSTICAL MODE

Let us calculate the ratio of the amplitudes A and B in the two cases. Let us see the situation for $K = 0$.

From eq. (*iii*) and (*v*), for $K = 0$, it follows that

$$\frac{A}{B} = -\frac{M_2}{M_1} \qquad \text{(for optical branch)}$$

and $$A = B. \qquad \text{(for acoustical branch)}$$

In the optical mode, the two types of atoms vibrate against each other with amplitudes in inverse ratio of their masses so that their centre of mass remains fixed. If the two atoms carry opposite charges, as in ionic crystals, then a vibration of this type can be excited by the electric field of a light wave. Hence the name optical branch.

In the acoustical mode, the two types of atoms (and their centre of mass) move together as happens in long-wavelength acoustical waves. Hence the name acoustical branch.

Quantisation of Lattice Vibrations—Phonons : The energy of a lattice vibration is quantised. The quantum of energy is called a "phonon" (in analogy with the photon of the electromagnetic wave). Thus, elastic waves in crystals are made up of phonons. Thermal vibrations in crystals are thermally-excited phonons (like the thermally-excited photons of black-body electromagnetic radiation).

The energy of each mode of vibration in an elastic wave of frequency v (or angular frequency $\omega = 2\pi v$) is

$$\varepsilon = (n + ½)\, hv = (n + ½)h\omega,$$

where $h = \frac{h}{2\pi}$ and n is the number of phonons in the mode, $h\omega$ is the quantum of elastic energy, that is, energy of a 'phonon' and ½ $h\omega$ is the zero-point energy of the mode.

The quantisation of lattice vibration energy has an important consequence when a lattice vibration is involved in an interaction. Since the energy of a mode of vibration has to be $(n+ ½)\, h_{\omega}$ with n any positive integer, and since to a first-order approximation the energy changes only to adjacent allowed values, we must have

$$\Delta\varepsilon = \pm\, h\omega$$

and correspondingly,

$$\Delta n = \pm 1.$$

This means that the energy change corresponds to a gain or loss of a single phonon at a time.

Experimental Evidence for Phonons : There is no direct experimental evidence that the energy of an elastic wave is quantised. (On the other hand, photoelectric effect is the direct experimental evidence that the energy of the electromagnetic wave is quantised). However, the following facts provide evidence for the existence of phonons :

(i) The lattice contribution to the heat capacity of solids approaches zero as the temperature approaches zero. This can be explained only if the lattice vibrations are quantised.

(ii) X-rays and neutrons are scattered inelastically by crystals. The change of energy and momentum in this process corresponds to the creation or absorption of one or more phonons. By measuring the energy and momentum of the scattered X-rays or neutrons we can determine the properties of phonons. Such experiments provide the strongest evidence for phonons.

Phonon Momentum

A phonon of wave vector $\vec{K}$ interacts with other particles such as photons, neutrons and electrons as if it has a momentum $h\vec{K}$. This is in accordance with de Broglie relation $p = \frac{h}{\lambda} = \frac{h/2\pi}{\lambda/2\pi} = h\vec{K}$). Sometimes $h\vec{K}$ is called the 'crystal momentum'.

A phonon on a lattice, however, does not carry physical momentum, except for $\vec{K} = 0$ which corresponds to a uniform translation of the crystal as a whole. But for most practical purposes a phonon acts as if its momentum were hK.

In crystals there exist momentum conservation laws (wave vector selection rules) for allowed transitions between quantum states. The elastic scattering (Bragg diffraction) of an X-ray photon by a crystal is governed by the wave vector selection rule

$$\vec{k} = \vec{k} + \vec{G}$$

where $\vec{G}$ is a vector in the reciprocal lattice, $\vec{k}$ is the wave vector of the incident photon and $\vec{k}$ is the wave vector of the scattered photon.

If the scattering of the photon is inelastic, with the creation of a phonon of wave vector $\vec{k}$, then the wave vector conservation law (selection rule) becomes

$$\vec{k}' + \vec{K} = \vec{k} + \vec{G}.$$

If a phonon $\vec{k}$ is absorbed in the process, then the relation would be

$$\vec{k}' = \vec{k} + \vec{K} + \vec{G}.$$

Specific Heat of Crystalline Solids : The heat capacity of a solid at constant volume is defined as

$$Cv = \frac{DE}{dr},$$

where E is the internal energy and T the Kelvin temperature.

The experimental facts about the heat capacity of solids in general are the following.

(i) At ordinary and high temperatures the value of the heat capacity of nearly all solids is close to $3Nk$, where N is Avogadro's number and k is Boltzmann's constant. For 1 mole of atoms it is 6 cal/mole-K (Dulong and Petit law).

Dulong and Petit regarded each atom of the solid as executing simple harmonic vibrations about its lattice site in three dimensions. Thus one mole of the solid, having N atoms, has $3N$ degrees of freedom. According to the classical law of equipartition of energy, each degree of freedom is assigned an average total (kinetic + potential) energy kT. Therefore, the total energy of one mole of the solid is

$$E = 3NkT = 3\ RT$$

where R is the universal gas constant. Thus, the heat capacity at constant Volume is

$$C_v = \frac{dE}{dT} = 3R = 6 \text{ cal/mole-K}.$$

(ii) At lower temperatures the heat capacity drops markedly and approaches zero. Near the absolute zero, the heat capacity varies as T^3 in insulators and as T in metals. If the metal becomes a superconductor, the drop is faster than T.

(iii) In magnetic solids there is a large contribution to the heat capacity near the temperature at which the magnetic moments become ordered. Below 0.1 K the ordering of nuclear moments may give very large heat capacities.

The first and second points are illustrated in the figure below for silicon and germanium.

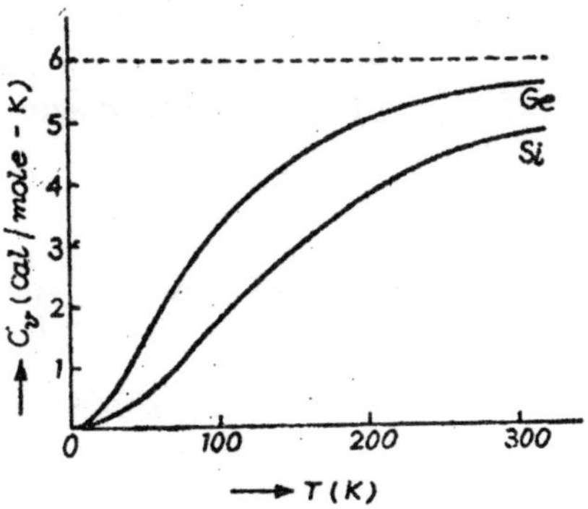

Einstein's Quantum Theory

Einstein, in 1907, applied quantum theory to the specific heat of solids. We know that the energy of a lattice vibration is quantised, the quanta of energy being 'phonons' which obey Bose-Einstein statistics. The energy of a phonon is $\hbar\omega$.

Einstein assumed that *all atoms of a solid lattice vibrate independently of each other with the same frequency in each of the three degrees of freedom.* Thus the energy of a lattice vibration depends only on its frequency ω (or phonon energy $\hbar\omega$) and phonon occupancy $<n>$, say. Now, one mole of the solid contains N atoms with $3N$ phonons of vibration, each with energy $h\hbar\omega$. Therefore, the energy per mole of the solid will be

$$E = 3N <n> \hbar\,\omega.$$

In thermal equilibrium at temperature T the occupancy is given by Bose-Einstein distribution

$$<n> = e^{(\epsilon-\epsilon F)/kT} \leq$$

and thus

$$E = {}_3N\ \hbar\,\omega\ \frac{1}{e^{\hbar\omega/kT}-1}$$

The specific heat is given by

$$C_v = \frac{dE}{dT} = 3N\ \hbar\,\omega\ \frac{e^{\hbar\omega/kT}\left(\frac{\hbar\omega}{kT^2}\right)}{\left(e^{\hbar\omega/kT}-1\right)^2}$$

$$= 3Nk\left(\frac{\hbar\omega}{kT}\right)^2 \frac{e^{\hbar\omega/kT}}{\left(e^{\hbar\omega/kT}-1\right)^2}$$

But $Nk = R$ (the universal gas constant).

$\therefore$
$$C_v = 3R\left(\frac{\hbar\omega}{kT}\right)^2 \frac{e^{\hbar\omega/kT}}{\left(e^{\hbar\omega/kT}-1\right)^2}.$$

Let us put $\frac{\hbar\omega}{k} = \Theta_E$, where Θ_E is known as Einstein characteristic temperature. Then we have

$$\boxed{C_v = 3R\left(\frac{\Theta E}{T}\right)^2 \frac{e^{\Theta E/T}}{\left(e^{\Theta E/T}-1\right)^2}}$$

This is Einstein's specific heat formula.

(i) At high temperatures ($T >> \Theta_E$) we have $\frac{\Theta_E}{T} << 1$, so that $e\Theta E/T \approx 1 + \frac{\Theta_E}{T}$ and we have

$$C_v \approx 3R\left(\frac{\Theta_E}{T}\right)^2 \frac{1+\frac{\Theta_E}{T}}{\frac{\Theta_E}{T}} \approx 3R$$

$$= 6 \text{ cal/mole-K.}$$

This is Dulong and Petit law which agrees with experiment; at high temperatures.

(ii) At low temperatures ($T << \Theta_E$) we have $\frac{\Theta_E}{T} >> 1$, so that $e^{\Theta_E/T} >> 1$, and the Einstein's formula gives

$$C_v \approx 3R\left(\frac{\Theta_E}{T}\right)^2 e^{-\Theta_E/T} \text{and we have}$$

This shows that as $T \to 0$; the specific heat C_v also approaches zero, a conclusion in general agreement with experiment.

The figure below compares the experimental values of C, (shown by dots) for diamond with the behaviour predicted by Einstein's theory (solid curve), using the characteristic temperature $\Theta_E = h\omega/k = 1320$ K. The agreement is reasonably good. (Θ_E or ω is an adjustable parameter and will have different values for different materials to match the experimental results.).

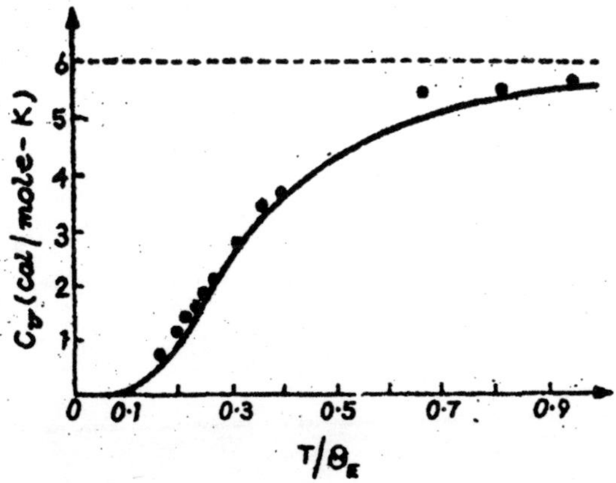

Limitation : There is no quantitative agreement at very low temperature. Near absolute zero, according to Einstein, $C_v \propto e^{-\Theta_E/T}$ whereas according to experiment, $C_v \propto T^3$. The reason is that Einstein oversimplified matters by assuming all of the phonons to have the same frequency.

In fact the phonon frequencies have a distribution like that in electromagnetic radiation in a cavity. This, in turn, is a result of the fact that the atomic vibrations in a lattice are strongly coupled and cannot be considered independent. The problem was taken up by Debye.

12

The Diodes

p-n Junction

Semiconductors, pure or doped (*p*-type or *n*-type), are bilateral; current flows in either direction with equal ease. If, however, in a semiconductor there exists a *p*-type region on one side and an *n*-type region on the other, then the semiconductor becomes unilateral; current flows easily in one direction only. The specific location in the semiconductor where the region changes from *p*-type to *n*-type (the lattice structure remaining continuous) is called the '*p-n* junction'. The semiconductor containing a *p-n* junction is called 'semiconductor diode'.

Fabrication of p-n Junction : A *p-n* junction is not the interface between two pieces of semiconductor of opposite types pressed together. It is a single piece of semiconductor crystal having an excess of donor impurities into one side and of acceptor impurities into the other. There are four important methods of making *p-n* junctions. These are growing, alloying, diffusing and ion implantation.

Growing Method : In order to grow a pure crystal of germanium (say), a "seed" crystal of germanium is touched to the surface of molten germanium and then slowly raised (Fig.). The molten germanium crystallises onto the seed and "grows" in the same crystalline structure as the seed. If the melt contains a donor impurity such as arsenic, an *n*-type semiconductor is obtained. When the *n*-type crystal has attained a reasonable size, an excess of an acceptor impurity, such as indium, is added to the melt. The remainder portion of the crystal is *p*-type. The boundary between the two materials is called a 'grown junction'. After the crystal is formed, it is cut into the desired shape. The grown junction has a rather gradual transition from the *n* to the *p* type.

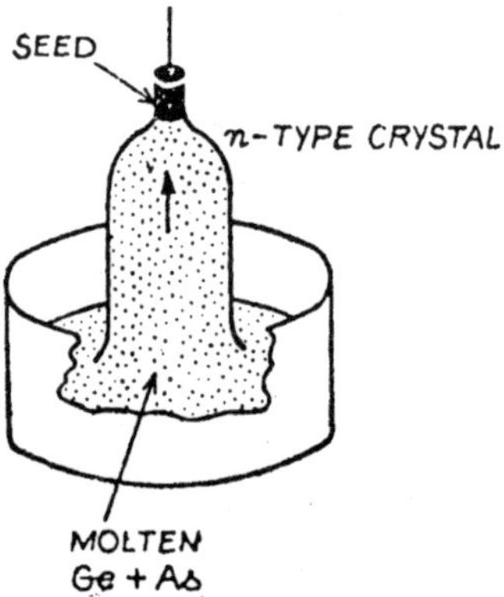

Alloying Method : In this method, a small pellet of an acceptor impurity, such as indium, is placed on an *n*-type semiconductor (say, germanium) wafer, as shown in figure below. The wafer is then put into a furnace in an inert atmosphere and the temperature is raised to the eutectic point. A thin film of In-Ge melt is formed on the interface. Upon cooling, the melt solidifies into *p*-type Ge, in continuation with the underlying *n*-type Ge. An 'alloy' *p*-*n* junction is thus formed between the re-crystallised region and the original wafer. In alloy junction the transition from *p*-type to *n*-type occurs in a small distance corresponding to a few atomic layers.

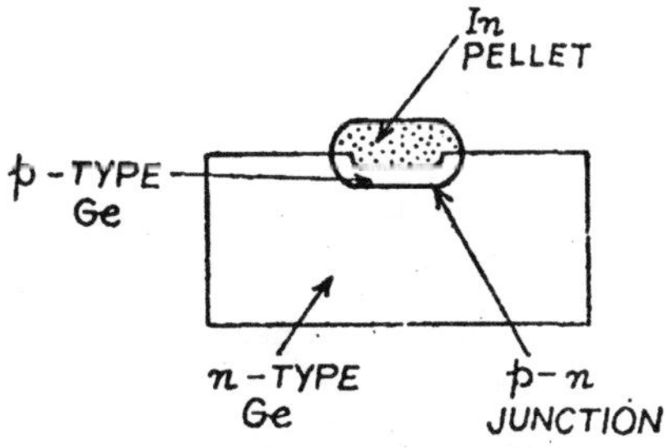

Diffusing Method : This is the most common way of making junctions today. In this method, first of all a thin layer of an inert oxide, such as silicon dioxide, is formed on the surface of an *n*-type semiconductor wafer by heating in oxygen. Then the oxide layer is removed from a selected area by etching and the wafer is put into a diffusion furnace at a temperature around 1000° C. Now, a *p*-type impurity in the gaseous form, such as boron gas, is introduced into the furnace. The acceptor atoms slowly diffuse into the *n*-type wafer through the selected area and a *p-n* junction is formed. (The untouched oxide layer is impervious). The concentration of impurity is maximum at the surface, gradually falling off towards the inside. By strict control of time and temperature, precise junction depth, usually a few microns, can be achieved. By selecting different areas on the surface of the wafer, one can make many junctions of small area on the same wafer.

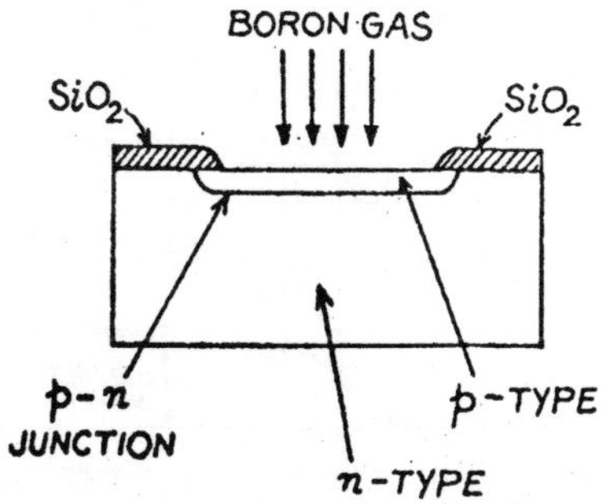

Ion Implantation Method : This is a new method in which impurity ions, accelerated to high energies, are shot at the semiconductor wafer. The ions penetrate the lattice to a depth of a fraction of a micron. The wafer is finally annealed at an elevated temperature.

Junction Diode : A junction diode is a semiconductor crystal having acceptor impurities (*p*-type) in one region and donor impurities (*n*-type) in another region. The boundary between the two regions is called '*p*-n junction'. The characteristic property of a *p-n* junction is that electric current can pass through it much more easily in one direction than in the other.

Potential Barrier at the Junction—Formation of Depletion Region : A *p-n* junction is shown in Fig. below. The *p*-type region has (positive) holes as majority charge-carriers, and an equal number of fixed negatively-charged acceptor ions. (The material as a whole is thus neutral). Similarly, the *n*-type region has (negative) electrons as majority charge carriers, and an equal number of fixed positively-charged donor ions.

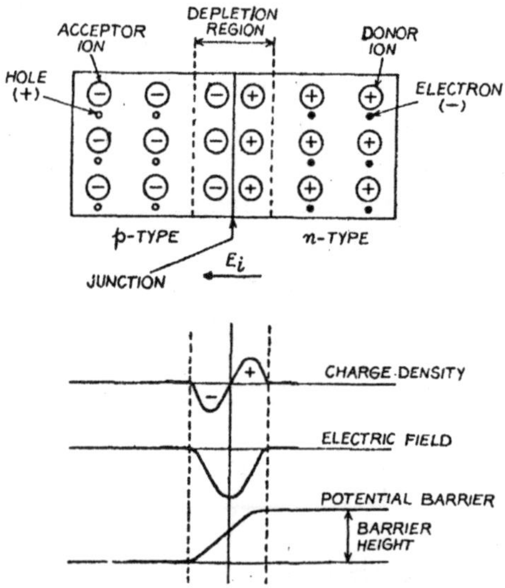

In addition to these majority charge-carriers, there are a few minority charge-carriers in each region. The *p*-region contains a few electrons while the *n*-region contains a few holes.

Thus, when the junction is formed, there is a charge-density gradient. This causes diffusion of charge-carriers across the junction. Holes diffuse from the *p*-region into the *n*-region, and electrons from the *n*-region into the *p*-egion. The diffused charge-carriers combine with their counterparts in the vicinity of the junction and neutralise each other.

The diffusion of holes leaves *uncovered* (negative) acceptor ions on the left of the junction and the diffusion of electrons leaves *uncovered* (positive) donor ions on the right. Thus, there is net negative charge on the *p*-side of the junction and net positive charge on the *n*-side. This sets up a potential difference across the junction, and hence an internal electric field E_i directed from the (positive) B-side to the (negative) *p*-side. Equilibrium is established when the field becomes large enough to stop further diffusion of the majority charge-carriers.

The field E_i however, helps the minority carriers (electrons in the *p*-region and holes in the *n*-region) to move across the junction.

The region on either side of the junction which becomes depleted (free) of the mobile charge-carriers is called the 'depletion region' (or 'transition' region). The thickness of this region is of the order of 10^{-6} m. The potential difference across the depletion region is called the 'potential barrier'. It is of the order of 0.1 to 0.5 volt depending on the temperature.

The general shapes of the charge distribution, the internal electric field and the potential barrier in the depletion-region are shown in the lower part of Fig. above.

Current Flow in Junction Diode (Rectifying Action) : In the absence of any external voltage applied across the *p-n* junction, there is no current in the diode. Under this condition, a few majority charge-carriers (holes in *p*-region and electrons in *n*-region) have sufficient energy to move across the junction in spite of the opposing internal field E_i and form a forward current. This current is, however, exactly balanced by the reverse current formed by the flow of minority carriers (electrons in *p*-region and boles in *n*-region) across the junction which is helped by the internal field. The net current is thus zero.

Forward Bias : When a battery is connected with its + pole joined to the *p-region* and — pole joined to the *n*-region, an external electric field E is set up directed from p towards n , in opposition to the small internal field E_i. As a result of the applied field E, holes in large number move from the *p*-region into the *n*-region and the electrons from the *n*-region into the *p*-region. These moving majority carriers form a relatively large forward current which increases with increase in the externally applied p.d. The small reverse current due to the flow of minority carriers remains unaffected. Thus, there is a net forward current. In this situation, the junction is said to be under 'forward bias'.

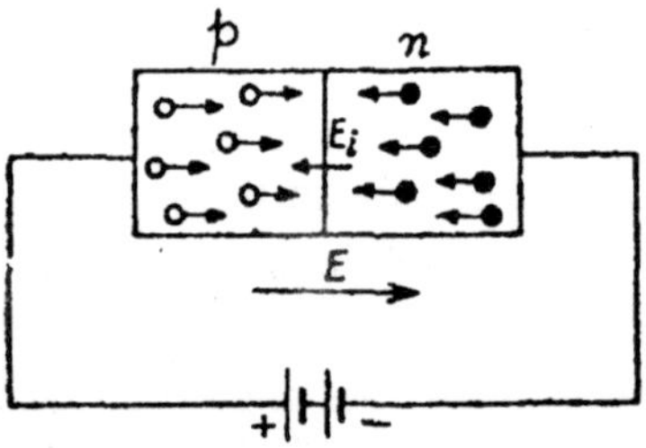

FORWARD BIAS

Reverse Bias : If, however, the battery is connected with its positive pole joined to the *n*-region and the negative pole joined to the *p*-region, the external field E is directed from n toward p, and thus aids the internal field E_i. The holes in the *p*-region and the electrons in the *n*-region are now pushed away from the junction and their motion stops. Hence, nominally, zero current results. In this situation the junction is said to be in the 'reverse bias'. The small reverse current due to the motion of minority carriers, however, exists. This current increases with increasing temperature.

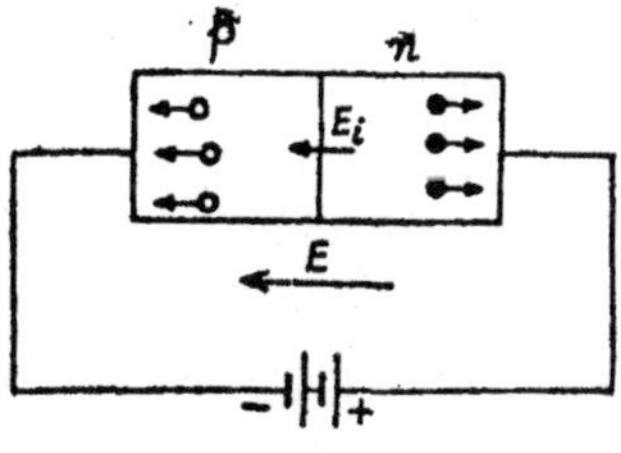

REVERSE BIAS

Thus we see that the junction diode offers a low resistance for the current to flow in one direction (under forward bias) but a very high resistance in the opposite direction (under reverse bias). It thus acts as a rectifier.

Characteristic (V-i) Curve of a Junction Diode : The *V-i* characteristic of a junction diode is shown in fig. below. With increasing forward bias the current increases rapidly, approximately exponentially. With reverse bias the (reverse) current remains very small over a long range, increasing very slightly with increasing bias.

Avalanche Breakdown : If the reverse bias is made very high (≈ 20 volts), the covalent bonds near the junction break down and a large number of electron-hole pairs are liberated. The reverse current then increases abruptly to a relatively large value. This is known as 'avalanche breakdown', and may damage the junction by the excessive heat generated unless the current is limited by external circuit.

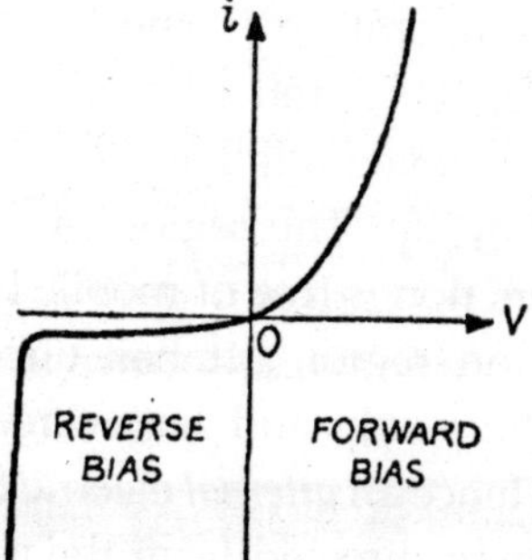

Energy Level Diagram for a p-n Junction : A *p-n* junction is the boundary between the *p*-type region (having acceptor impurities) and the *n*-type region (having donor impurities) of a doped semiconductor. The energy level diagram of the *p-n* junction may be visualised most easily by considering the *p*-type and *n*-type semiconductors as being initially separated.

The Fermi energy Î$_F$ is closer to the top of the valence band in the *p*-type material, and closer to the bottom of the conduction band in the *n*-type. When the junction between the two types of material is formed, holes flow from the *p*-type region into the *n*-type region via the valence band ; while electrons flow in the opposite direction via the conduction band. As a result of this flow, *equilibrium is established when the Fermi levels in the two regions are along the same line.*

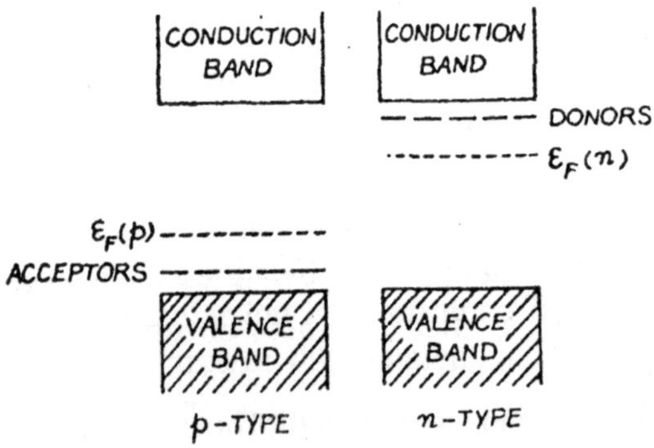

The resulting energy level diagram of an unbiased *p-n* junction at room temperature is shown in Fig. below. The conduction bands of the *p* and *n* regions, and similarly the valence bands, have moved relatively so that the Fermi-levels ϵ_F (*p*) and ϵ_F (*n*) line up. The region in the vicinity and on either side of the junction is free of mobile holes and electrons and is the 'depletion region'. It has uncovered (negative) acceptor ions on the *p*-side and uncovered (positive) donor ions on the *n*-side. Hence an *internal* electric field E_i is produced in the depletion region directed from the *n*-side to the *p*-side.

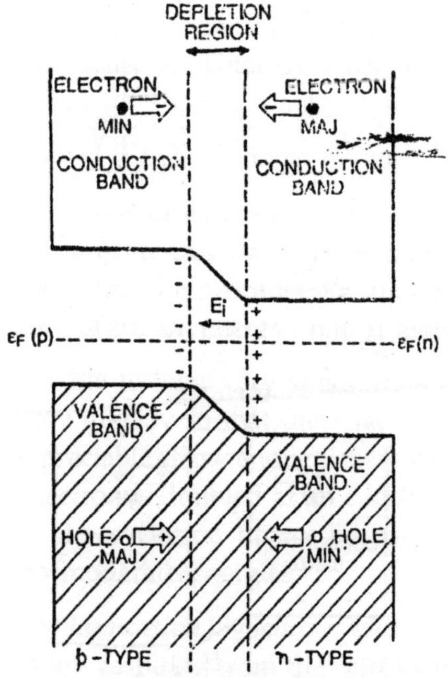

UNBIASED

Even after equilibrium is established, there is still a flow of holes and electrons across the junction. A few of the majority, carriers (holes in *p*-region and electrons in *n*-region) are able to overcome the internal field E_4 (which opposes them) and form a toward current. At the same time, the minority carriers (electrons in *p*-region and holes in *n*-region), which are aided by the field *Ei*, move across the junction and form a reverse current. These two small currents neutralise each other so that the net current is zero. The movements of holes and electrons are shown by arrows in Fig. above.

Forward Biased Junction : The junction is given a forward bias by connecting across it an external voltage source such that positive voltage is applied to the *p*-region and negative voltage to the *n*-region. This creates an external field *E* opposite to internal field E_i, and raises the electron energy levels of the *n*-type semiconductor- relative to those of the *p*-type semiconductor. The Fermi level in the n-region is raised relative to that in the p-region by eV where V is the applied voltage. The electrons in the n-region now easily cross over to the p-region; and on a similar argument, the holes in the p-region easily cross over to the n-region. The majority carriers in each of the two regions thus produce a large forward current. The reverse current due to minority carriers remains unaffected by the applied voltage. The equilibrium of currents flowing in opposite directions is therefore upset, and the contribution of majority carriers to the net current dominates.

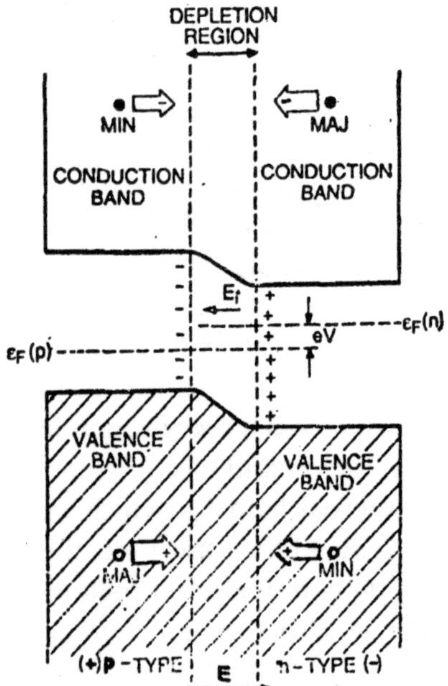

FORWARD BIASED

Reverse Biased Junction : When the positive voltage is applied to the n-region and negative voltage to *p-region,* the junction is said to be reverse biased. The external field E is now in the same direction as the internal field Et ; the Fermi level in the n-region moves down with respect to that in the p-region. The majority carriers in the two regions now hardly produce any current. The equilibrium of currents is again upset, and the small current due to minority carriers, which is not much affected, dominates.

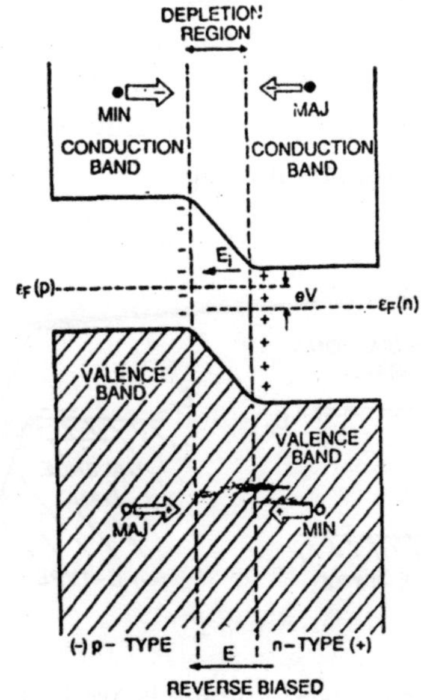

Zener Diode

In an ordinary semiconductor diode under *reverse* bias, the (reverse) current remains almost constant over quite a long voltage-range, until 'avalanche breakdown' occurs when the (reverse) voltage becomes ≈ 20 volts. The current then increases abruptly due to a sudden increase in the number of electron-hole pairs. If, however, both the *p*- and *n*-type regions of the diode are heavily doped, the reverse current may increase abruptly at a (reverse) voltage lower than that required for the avalanche breakdown.

The breakdown in this case is known as 'Zener breakdown' and arises from the quantum mechanical potential barrier penetration (tunneling) across the depletion region. Such a semiconductor device is called a 'Zener diode'.

The energy band structure of an unbiased heavily-doped *p-n* junction diode is shown in Fig. below. As a result of an increased number of donors and acceptors, the depletion region is narrower. ' Moreover, the Fermi level $\in_F$ in the *p*-type region is closer to the valence band and that in the *n*-type region is closer to the conduction band, compared to a light-doped junction. However, under equilibrium condition, the Fermi levels in the two regions line up.

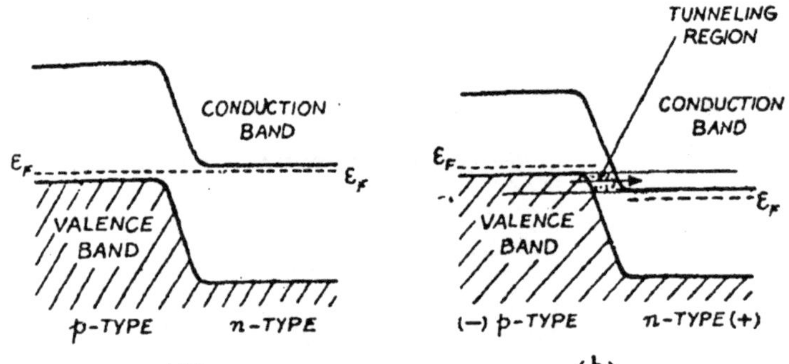

(a) (b)

When a reverse-biased voltage is applied to the diode, the energy levels in the *n*-type semiconductor moves down relative to those in the *p*-type semiconductor. At a certain (reverse) voltage, called the Zener voltage, the bottom of the conduction band in the *n*-region becomes lower than the top of the valence band in the *p*-region (Fig. above). Electrons now "tunnel" directly across the potential barrier from the valence band in the *p*-region into the conduction band in the *n*-region.

Since the barrier (depletion region) is narrow, the tunneling probability is high and a large (reverse) current results. Zener breakdown can occur in heavily doped diodes at 6 volts or less.

The resulting reverse characteristic of a Zener diode is shown in Fig. below. Because of the voltage limiting characteristic of a reverse-biased Zener diode it is possible to use these diodes as voltage stabilisers.

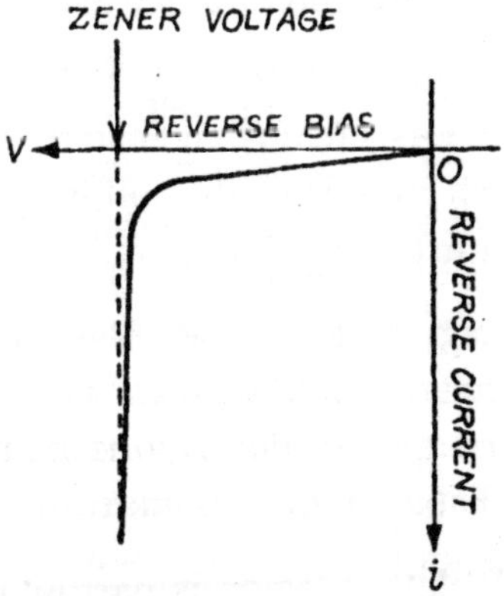

Tunnel Diode

A tunnel diode is a semiconductor device which makes use of the quantum mechanical phenomenon of potential barrier penetration. It is a p - n junction made from *quite heavily doped* semiconductors. The energy band structure of such a junction, when unbiased, is shown in Fig. below. It differs from the band structure of an ordinary p-n junction in two respects :

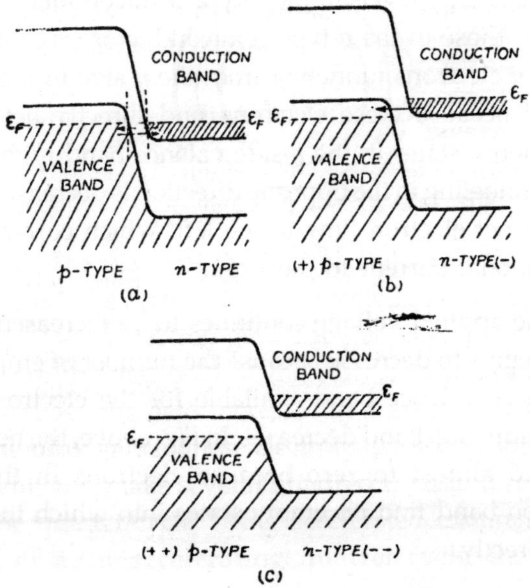

(i) The depletion region is much narrower $\approx 10^{-8}$ m,

(ii) The donor and acceptor levels, in the *n*-type and *p*-type semiconductors, are no longer sharp but become broad bands which merge into the bottom of the *n*-conduction band and into the top of the *p*-valence band respectively.

The Fermi energy level $\in_F$, therefore, moves up into the conduction band on the *n*-side and down into the valence band on the *p*-side. When the junction is formed, however, the two Fermi levels match under the equilibrium condition.

Because the depletion region is narrow ($\approx$ a few electron wavelength), electrons can "tunnel" directly through the forbidden band there, from the conduction band of the *n*-side into the valence band of the *p*-side, and vice-versa. When no external voltage is applied across the junction, the rate of electron tunneling is same in both directions and the net current is zero.

When a small forward-biased voltage is applied to the diode, the energy levels in the *n*-type semiconductor move up relative to those in the *p*-type semiconductor (Fig.). Now, the number of electrons tunneling from the *n*-side to the *p*-side is increased because these electrons find directly before them empty energy states in the *p*-side valence band ; whereas the number tunneling in the opposite direction is decreased. Hence, there is a net electron flow to the left, which corresponds to a conventional current to the right.

As the applied voltage continues to be increased, the net current begins to decrease because the number of empty states in the *p-side* valence band available for the electrons in the *n*-side conduction band decreases. In Fig. above, the net current is reduced almost to zero because electrons in the *n*-side conduction band find no empty states into which they could tunnel directly.

With still higher applied voltage, the current becomes that characteristic of an ordinary *p-n* junction diode. The net current is now due to the 'diffusion' of majority charge-carriers (electrons in the conduction band on the *n*-side and holes in the valence band on the *p*-side) across the junction.

Fig. below shows the voltage-current characteristic curve of a tunnel diode. The points *a, b* and c correspond to the parts *a, b* and *c* of the previous figure. (The dashed curve indicates the behaviour of an ordinary junction diode). In the region between points *b* and *c* the tunnel diode has a *negative* resistance, the current decreasing with increasing voltage. This feature makes it specially useful in the switching circuits of computers.

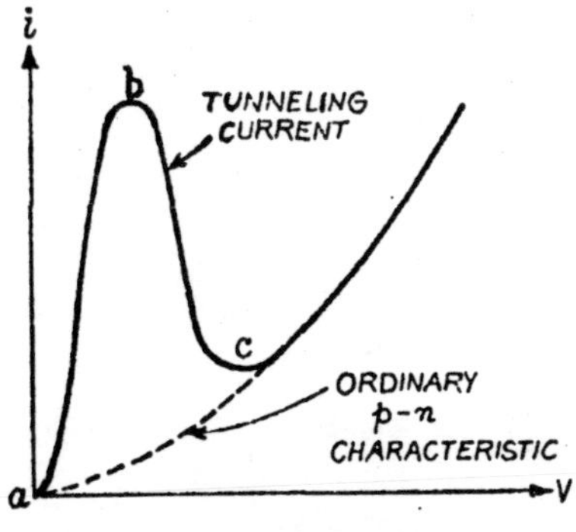

Applications : The greatest advantage of the tunnel diode is its very fast response (change in current) to the voltage-change in the region *a* to *c*. In ordinary diodes and transistors, the response time depends on the diffusion speed of the charge-carriers which is low. Hence these devices cannot be used at high frequencies. Tunnel diodes, on the other hand, can be used as oscillators at frequencies above 10^{11} Hz, and in switching circuits operating in time-intervals $\approx 10^{-9}$ sec.

13

The Semiconductors

The semiconductors are the solids whose electrical conductivity lies between the very high conductivity of metals and the very low conductivity of insulators. They are characterised by a *narrow* energy gap $\in g$ (~ 1 eV) between the valence band and the conduction band (Fig.). The most commonly known semiconductors are germanium and silicon which have energy gaps of about 07 and 11 eV respectively.

At absolute zero *(T = 0)*, *all* states (energy level) in the valence band are full and all states in the conduction band are empty. An applied electric field cannot give so much energy to the valence electrons that they could cross the gap $\in_g$ and enter the conduction band. Hence *pure semiconductors are insulators at low temperatures.*

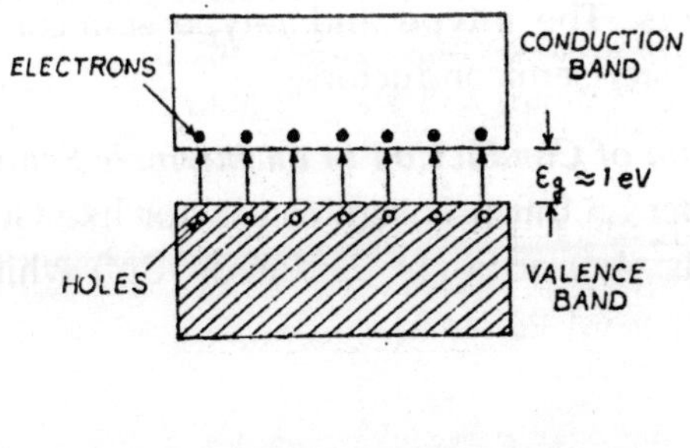

At room temperature, however, some of the valence electrons acquire thermal energy greater than $\in g$ and cross over into the conduction band (Fig.). These free electrons, and the (positive) holes created in the valence band, can move about even under a small applied field. The solid is therefore slightly conducting, that is, it is a semiconductor. (The conductivity further increases with increasing temperature).

Solids having a *partially-filled* upper band, or *overlapping* valence and conduction bands, are conductors, because electrons in this band can easily acquire additional energy from an applied electric field and move to the higher energy levels in the same band. On the other hand, solids having a wide energy gap ($\in g \approx 10$ eV) between a filled valence band and an empty conduction band are insulators, because valence electrons cannot acquire so much energy from an applied field that they could cross the gap and enter the conduction band. Hence conduction in them is impossible.

Types of Semiconductors

Pure germanium and silicon are 'intrinsic' semiconductors, and their conductivity (which arise from thermal excitation) is intrinsic, conductivity. If a small amount of some impurity be introduced into a pure semiconductor, its conductivity increases appreciably. Such impure semiconductors are called 'extrinsic or impurity semiconductors, and the resulting conductivity is called' extrinsic conductivity...,

The process of introducing the impurity is known as doping. Hence the impurity semiconductors are also known as doped semiconductors. The *n*-type and *p*-type semiconductors are doped (extrinsic) semiconductors.

***Mechanism of Conduction in an Intrinsic Semiconductor*:** Let us consider an intrinsic semiconductor like Ge ($Is^2\, 2s^2\, 2p^6\, 3s^2\, 3p^6\, 3d^{10}\, 4s^2\, 4p^2$) or Si ($1s^2\, 2s^2\, 2p^6\, 3s^2\, 3p^2$) which are both tetravalent.

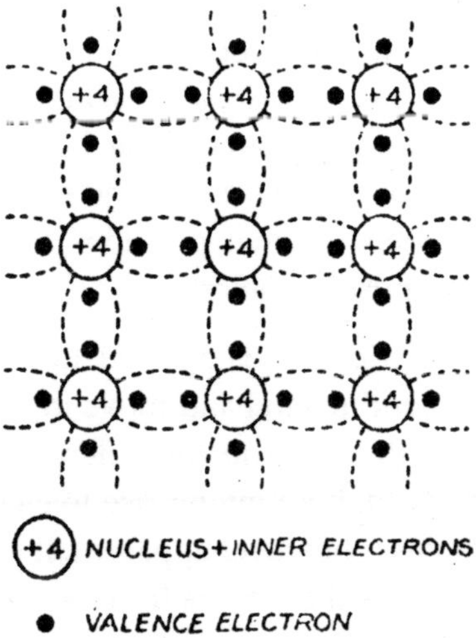

An atom of Ge (or Si) has four valence electrons surrounding an inert core (nucleus + tightly-bound inner electrons) having a charge of + 4e. The crystal structure of these materials consists of a three-dimensional tetrahedral pattern with an atom at each vertex. This is illustrated in two dimensions in figure above. Each of the four valence electrons of an atom is shared by one of its four nearest neighbouring atoms. The covalent bonds so formed, which are shown by dashed lines, provide the binding forces between the neighbouring atoms.

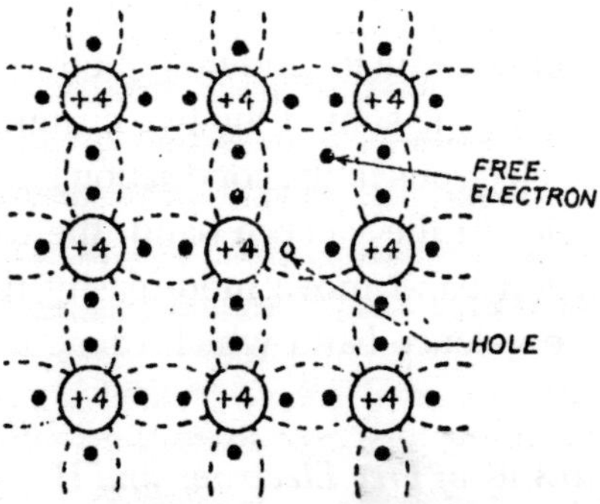

At temperatures close to absolute zero, all valence electrons are tightly bound to the core (all are said to be in the valence band), and so no free electrons are available to conduct electricity through the crystal. However, at room temperature, a few of the valence electrons are *thermally* excited into the conduction band (that is, a few of the covalent bonds are broken) and become free to move about (Fig.). The excited electrons leave "vacancies" at their original positions in the valence band which serve as "positive holes".

When an electric field is applied to the crystal, the free electrons in the conduction band move in a direction opposite to the field, and the holes in the valence band move in the direction of the field, both giving rise to electric currents. The motion of the holes is however apparent. It is in fact the motion of the otherwise bound electrons from one vacancy to the next *within the valence band* in a direction opposite to the applied field.

Thus, in a semiconductor electrons and holes both constitute current. The total current is the sum of the two currents. Although the number of free electrons in the conduction band is equal to the number of holes in the valence band, the electron contribution to the current is larger. This is because the (free) electrons in the conduction band move more easily than the (bound) electrons in the valence band which cause the motion of the holes.

Thus in a semiconductor and holes both constitute current. The total current is the sum of the two currents. Although the number of free electrons in the conduction band is equal to the number of holes in the valence band, the electron electrons in the conduction band move more easily than the (bound) electrons in the valence band which cause the motion of the holes.

Concentrations of Free Electrons and Holes in an Intrinsic Semiconductor : The electrons in the conduction, band, and the holes in the valence band, of an intrinsic semiconductor are governed by Fermi-Dirac distribution law. We may, therefore, write for the number of electrons per unit volume, having energies between $\in$ and $\in + d\in$, in the conduction band

$$n(\in)\, d\in = g(\in) f(\in), \qquad \text{... } (i)$$

where g $(\in)\, d\in$ is the number of quantum states (energy levels) available to these electrons in the conduction band and $f(\in)$ is the probability of an electron occupying a given energy level $\in$. The function $f(\in)$ is known as the 'Fermi function' and is given by

$$f(\in) = \frac{1}{e^{(\in - \in_F)/kT} + 1} \qquad \text{... } (ii)$$

$\in_F$ is the 'Fermi energy' measured upwards from the top of the valence band as shown in Fig. Since there are no levels in the band gap $0<\in<\in g$, we have

$$g(\in) = 0 \qquad (0<\in<\in_g),$$

where $\in = 0$ denotes the top of the valence band and $\in g$, is the energy gap.

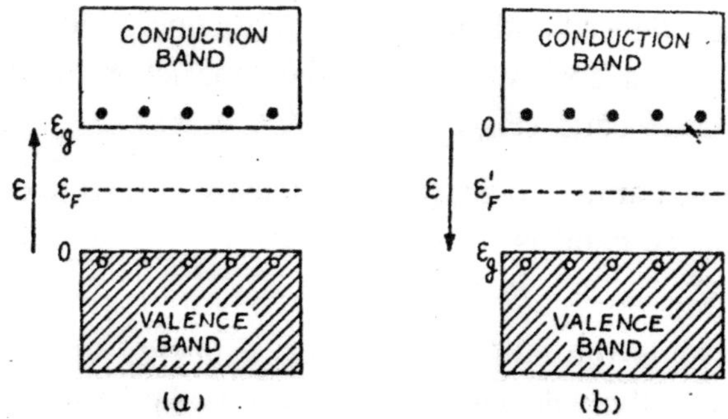

If the conduction band is considered extended from E_g to ∞, then the density of electrons in the conduction band is

$$n_e = \int_{\in_g}^{\infty} g(\in)\, d\in f(\in)$$

$$= \int_{\in_g}^{\infty} \frac{g(\in)\, d\in}{e^{(\in - \in_F)/kT} + 1}$$

In the conduction band, $\in - \in_F > kT$, so that the exponential term is very much greater than 1. Therefore, the last equation reduces to

$$n_e = \int_{\in_g}^{\infty} g(\in)\, e^{-(\in - \in F)/kT}\, d\in$$

$$= \left| g \in e^{-(\in - \in F)/kT} (-kT) \right|_{\in_g}^{\infty}$$

$$= g(\in_g)\, kT\, e^{-(\in_g - \in_F)/kT}$$

$$= K_x\, e^{-(\in_g - \in_F)/kT} \qquad \text{... (iii)}$$

where K_e is a temperature-dependent term and can be taken as the effective number of energy levels per unit volume in the conduction band.

We can analyse the holes in the valence band in the same manner as electrons in the conduction band. The positive holes may be considered as being excited *from* the conduction band *into* the valence band, and so the Fermi energy for them must be measured downwards from the conduction band towards the valence band (Fig.).

Thus, parallel to eq. (*iii*), the density of (positive) holes in the valence band is

$$n_h = K_h\, e^{-(\in_g - \in'_F)/kT}, \qquad \text{... (iv)}$$

where K_h is the effective number of energy levels per unit volume in the valence band, and $\in' F$ is the Fermi energy for holes.

Since, in the intrinsic semiconductor, the conduction electrons and positive holes are produced in pairs, we have

$$n_e = n_h$$

and so eq. (*iii*) and (*iv*) require that

$$K_e\, e^{-(\epsilon g - \epsilon F)/kT} = K_h\, e^{-(\epsilon g - \epsilon' F)/kT}.$$

If the energy level distributions are approximated to be the same for the conduction and valence bands, then

$$K_e = K_h,$$

and we must have

$$\epsilon_g - \epsilon_F = \epsilon_g - \epsilon'_F$$

or $$\epsilon F = \epsilon' F\ .$$

This means that, *the Fermi level lies in the centre of the energy gap*, and

$$\epsilon F = \epsilon' F = ½\ \epsilon_g\ .$$

The equations (*iii*) and (*iv*) can now be written as

$$n_e \simeq K_{,e}{}^{-\epsilon_g/2kT}$$

and $$n_h \simeq K_h\ {}^{e-\epsilon_g/2kT}$$

for the intrinsic semiconductor. Although the terms K_e and K_h are temperature-dependent, each being proportional to $T^{3/2}$, but the variation in n_e and n_h is mostly dependent on the exponential term involving — $\epsilon g/2kT$.

Change in Conductivity with Temperature

In an intrinsic semiconductor, both the electrons and the holes contribute to the conductivity. Hence the total conductivity can be written in terms of the densities of both charge-carriers and their mobilities, as

$$\sigma = n_e\, e\, \mu_e + n_h\, e\, \mu_h\ ,$$

where μ_e and μ_h, are the mobilities of electrons and holes respectively and *e* is electronic charge. For the pure (intrinsic) semiconductor, $n_e = n_h = ni$, where n_t is the intrinsic concentration.

$$\therefore \qquad \sigma = ni\ e\ (\mu_e + \mu_h)$$

Since the mobility varies relatively slightly with temperature, the dependence of conductivity on temperature is almost the same as that of *ni* . We know that

$$n_i \propto e^{-\epsilon_g/2kT}$$

where ∈g is the energy gap. Therefore, we can write

$$\sigma = Ke^{-\epsilon_g/2kT},$$

where K is some constant. This shows that *as the temperature increases, the conductivity of the semiconductor increases exponentially.* (This is because the density of electron-hole pairs increases). If ∈ ∞. be the (extrapolated) conductivity for $T = \infty$, then

$$\sigma\,\infty = K$$

and we write

$$\sigma = \sigma_\infty\, e^{-\epsilon_g/2kT}$$

or $\log_e$ $$\sigma = \log_e \sigma\,\infty\ \frac{\epsilon_g}{2kT}$$

A plot between log, σ and $1/T$ gives a straight line whose slope is — ∈g/2*k* (fig.). This is one of the methods, of determining the value of the semiconductor energy gap.

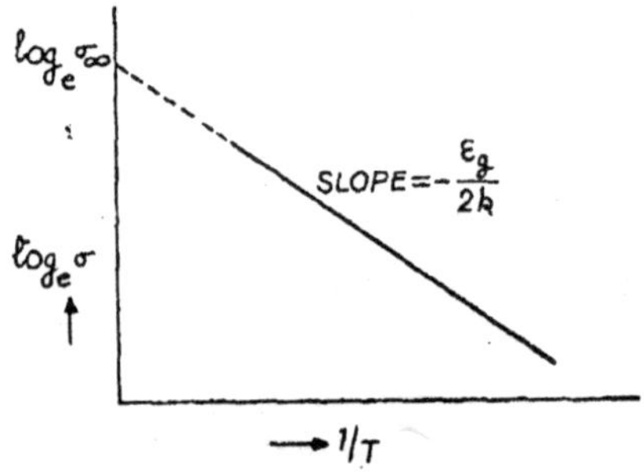

The substantial increase in the conductivity of semiconductors with increase in temperature is utilised in "thermistors" which find extensive application in thermometry, in the measurement of microwave-frequency power, as a thermal relay, and in controlling devices based on temperature-changes.

In contrast, the conductivity of a metal decreases almost linearly with increasing temperature. This is because the increase in temperature results in greater lattice-ion vibrations and hence decrease in mean free path λ of the free electrons. The result is a decrease in the mobility, and hence in conductivity $\left(\sigma = \frac{n_e e^2 \lambda}{2mv_F} = n_e e \mu_e, \text{ where } \mu_e = \frac{e\lambda}{2mv_F}\right)$ In a semiconductor also, the mobility decreases, but the increase in the number of electron-hole pairs (charge-carriers) with increase in temperature overwhelms the mobility effect.

In a metal, on the other band, there is no significant increase in the number of charge-carriers with increase in temperature.

Extrinsic (or Impurity) Semiconductor : If a small amount of a pentavalent or trivalent impurity is introduced into a pure germanium (or silicon) crystal, the conductivity of the crystal increases appreciably, and the crystal becomes an 'extrinsic' semiconductor.

Extrinsic semiconductors are of two types : n-type and p-type n-type Semiconductors : When a pentavalent (antimony, phosphorus or arsenic) atom replaces a Ge (or Si) atom in the crystal lattice (Fig.), four of its five valence electrons form covalent bonds with the neighbouring Ge (or Si) atoms. The fifth electron requires lattice energy to be detached from the impurity atom (only 0.01 eV for Ge and 005 eV for Si).

It therefore becomes free at room temperature ($kT \sim 0025$ eV) to move about in the crystal and acts as a charge-carrier. The crystal is now called an 'n-type' semiconductor because it has an excess of 'negative' charge-carriers (electrons). The impurity is called a 'donor' because it *donates* the conducting electrons.

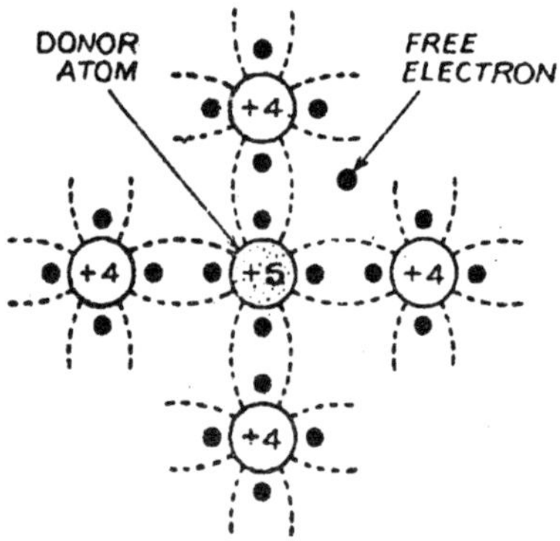

On an energy-level diagram, the impurity atoms introduce discrete energy levels for the electrons just below the conduction band (Fig.). These are called 'donor impurity levels'. They are only 0.01 eV below the conduction band in case of Ge (0.05 eV in case of Si). Therefore, at room temperature, the "fifth" electrons of almost all the donor atoms are thermally excited from the donor levels into the conduction band where they move as charge-carriers when an external field is applied.

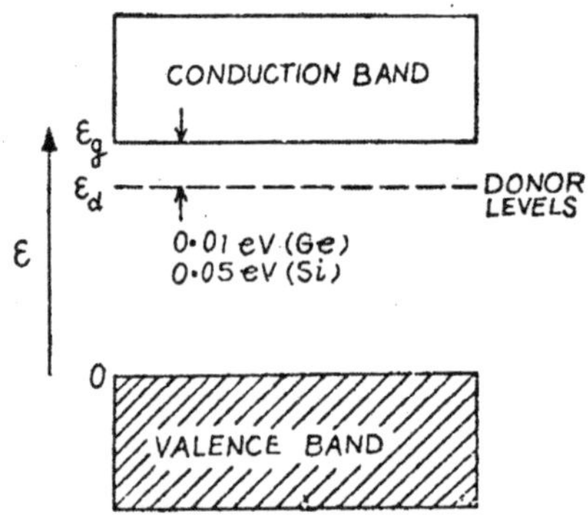

At ordinary temperature, almost all the electrons in the conduction band come from the donor levels, only a few come from the valence band. Therefore, the main charge-carriers responsible for conduction are the electrons contributed by the donors. Since the excitation from the valence band is small, there are very few holes in this band. The current contribution of the holes is therefore small. Thus, in an *n*-type semiconductor the electrons are the 'majority carriers', and the holes are the 'minority carriers'.

***p*-Type Semiconductors :** We can alternatively introduce a trivalent impurity into an intrinsic semiconductor. When a trivalent (boron, aluminium, gallium or indium) atom replaces a Ge (or Si) atom in the crystal lattice (Fig.), only three valence electrons are available to form covalent bonds with the neighbouring Ge (or Si) atoms. This results into an empty space or a positive "hole" around the impurity atom. An electron bound to a neighbouring Ge (or Si) atom drops into this hole, when an external field is applied, thereby creating a new hole. This phenomenon continues and the hole moves in the crystal lattice, acting as a positive charge-carrier. The crystal is called a '*p*-type' semiconductor because it has an excess of positive charge-carriers (holes). The impurity is called an 'acceptor' because the impurity atoms create holes which accept electrons. (Fig.)

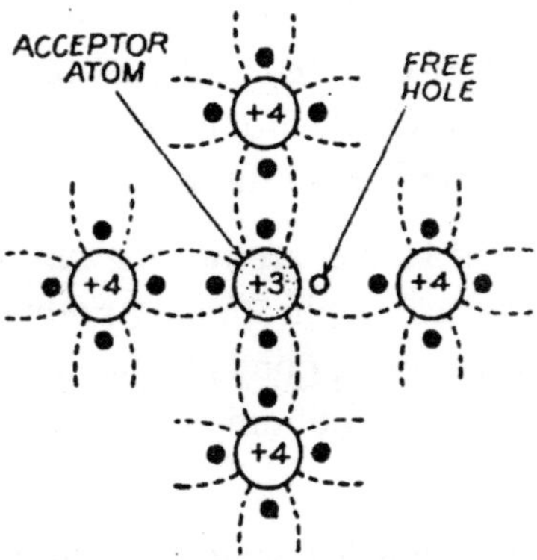

In the energy-level diagram, the trivalent impurity atoms introduce vacant discrete levels just above the top of the valence band (Fig.). These are called 'acceptor impurity levels'. At room temperature, electrons are easily excited from the valence band into the acceptor levels. The corresponding holes created in the valence band are the main charge-carriers in the crystal when an electric field is applied. Thus, in a p-type semiconductor the holes are the 'majority carriers' ; and the few electrons, thermally excited from the valence band into the conduction band, are 'minority carriers'.

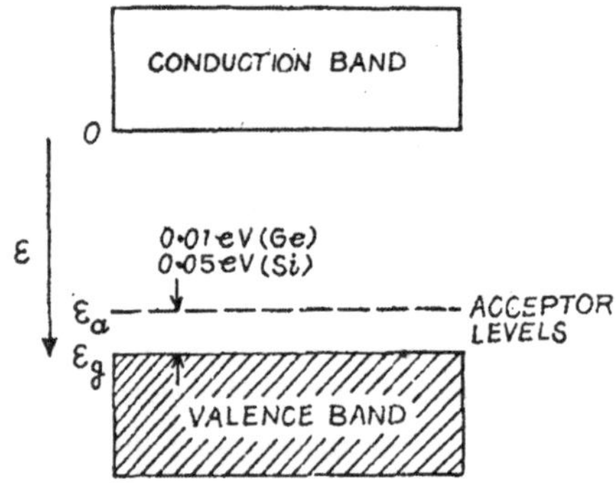

The conductivity of an intrinsic semiconductor is very poor unless the temperature is very high. At room temperature only one atom in 10^9 contributes to the conduction. The intrinsic semiconductor is thus of no practical use. In an extrinsic semiconductor, however, any desired conductivity can be precisely achieved by controlling the impurity concentration. If one impurity atom is added for every 10^8 germanium atoms, the conductivity increases 16 times.

Germanium and silicon are not the only semiconducting materials with practical applications. Another important class of semiconductors has compounds of trivalent and pentavalent elements, such as Ga As (gallium arsenide) and In Sb (indium antimonide).

Fermi Level in Impurity Semiconductors : In an intrinsic (pure) semiconductor, the Fermi level $\in F$ lies in the middle of the energy gap between the valence band and the conduction band, indicating equal concentrations of free electrons and holes.

The situation is different for impurity semiconductors. In an n-type semiconductor, the Fermi level lies above the middle of the energy gap because there are more electrons in the conduction band than there are holes in the valence band. On the other hand, in a *p*-type semiconductor, the Fermi level lies below the middle of the energy gap because there are fewer electrons in the conduction band than there are holes in the valence band. The position of the Fermi level, however, changes with impurity concentration as well as with temperature.

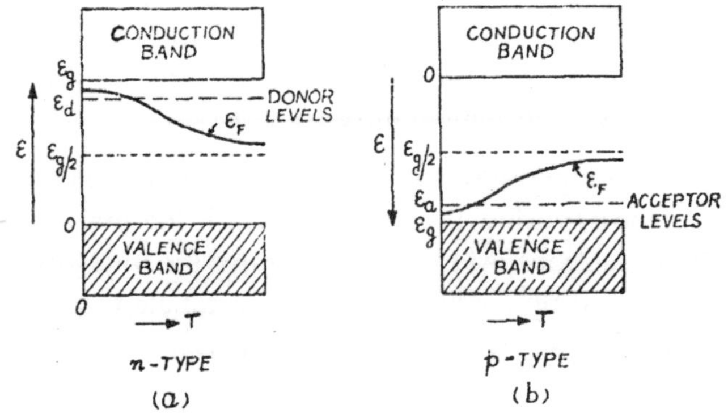

In an *n*-type semiconductor, at $T = 0$, the donor levels are all occupied but there are no electrons in the conduction band. Therefore, the Fermi level $\in_F$ lies between the donor levels and the bottom of the conduction band (all states below $\in_F$ are full and all above $\in_F$ are empty), as indicated in Fig. (a). Now, as the temperature is increased above absolute zero, electrons are raised from donor levels to the conduction band, and the Fermi level moves down.

At the temperature at which half of the donor states have become empty, the Fermi level coincides with the donor levels. With further increase in temperature, the thermal excitation of electrons from the valence band to the conduction band starts (as in an intrinsic semiconductor) and the Fermi level drops more toward the centre of the energy gap. When the number of these thermally-generated electrons (from the valence band) in the conduction band becomes much larger than the donor electrons there, the Fermi level drops to nearly the centre of the energy gap (Fig. a). Under these conditions (low donor concentration and high temperature) the concentration of electrons in the conduction band and that of holes in the valence band become almost equal and the crystal behaves like an intrinsic semiconductor

The same type of argument for the *p*-type semiconductor leads to the conclusion that, as the temperature is increased, the Fermi level moves from between the top of the valence band and the acceptor levels, at $T = 0$, to the centre of the energy gap at high temperatures (Fig. b).

At low temperatures, where $kT < \in_g$, conduction is due mostly to the impurities because there is little thermal excitation of valence electrons. At high temperatures the impurity levels have been used up (*i.e.* they have either donated or accepted electrons), and the semiconductor conducts as though it were intrinsic.

Various Concentrations in Semiconductors

A semiconductor contains free electrons and holes which are generated in pairs due to thermal excitation. The rate of thermal generation (electron-hole pairs/m^3-sec) depends upon the properties of the material of the semiconductor and is a function of temperature. Along with their generation, the electrons and holes continue to recombine and disappear. The rate of recombination depends upon the numbers of electrons and holes present. If they are few in number, the rate of recombination is low ; if they are many, the rate is high. In general, the rate of recombination is given by

$$R = r\, n_e\, n_h ,$$

where n_e and n_h are electron and hole concentrations (number of electrons or holes/m^3) respectively and r is a proportionality constant for the material.

Under equilibrium conditions, the rate of generation, g (say), is just equal to the rate of recombination, that is,

$$g = R = r\, n_e\, n_h\,.$$

In a pure (intrinsic) semiconductor the concentrations of electrons and holes are equal, that is,

$$n_e = n_h = n_i,$$

where ni is called the intrinsic concentration. Then, we have

$$g = R = r\, n_e\, n_h = r\, n_i^2.$$

Even in a doped (impurity mixed) semiconductor, the bulk of the atoms are still germanium (or silicon) and the thermal generation rate is unchanged from the intrinsic value. Therefore, the last equation is a general relation at a given temperature. It gives

$$n_e\, n_h = n_i^2\,,$$

for doped semiconductors as well. This relation is called 'mass-action' law.

The electron and hole densities are further inter-related because of the fact that the semiconductor as a whole is electrically neutral.

Let N_d be the concentration of donor atoms in an n-type semiconductor. These are practically *all* ionised (activated), giving N_d positive ions per m^3 which are fixed in the lattice. The free positive-charge (hole) density in the semiconductor is n_h. Hence the *total* positive-charge density is $N_d + n_h$. In n-type semiconductor, the total negative-charge density is just the electron density n_e. For electrical neutrality of the semiconductor, we must have

$$ne = N_d + n_h.$$

In practice, in an n-type semiconductor the hole concentration is very small compared to electron concentration.

$$\therefore \qquad n_e \simeq N_d\,.$$

Thus, *in an n-type semiconductor the free-electron concentration is approximately equal to the density of donor atoms.* The concentration of holes in the n-type semiconductor is

$$n_h = \frac{n_i^2}{n_e} \approx \frac{n_i^2}{n_d}$$

The corresponding equations for a p-type semiconductor are

$$n_h \simeq N_a$$

and $$n_e \simeq \frac{n_i^2}{n_a},$$

where N_a is the density of acceptor atoms.

PROBLEMS

1. ***Compare the number of conduction electrons is a semiconductor, a conductor and an insulator.***

Ans. Assuming the Fermi energy to lie in the middle of the energy gap, the occupation probability of conduction band at temperature T is

$$e^{-\epsilon_g/2kT}$$

where $\in_g$ is the energy gap.

At room temperature (T = 300K), we have

$$kT = (1.38 \times 10^{-23} \text{ joule}/K)\ (300\ K)$$

$$= \frac{4.14 \times 1-^{-21}\ joule}{1.60 \times 10^{-19}\ joule/eV} \simeq \frac{1}{40} \text{ eV}.$$

For a semiconductor, $\in_g \simeq 1$ eV, therefore

occupation probability $\simeq e^{-20} \approx 10^{-9}$,

that is, 1 atom in 10^9 contributes an electron to the electrical conductivity.

In a metal essentially every atom contributes an electron to the conductivity, that is

occupation probability ≈ 1.

An insulator has a band structure very similar to that of a semiconductor, except that the energy gap is, say, 5 eV instead of 1 eV. This small difference in the width of the energy gap has an enormous effect on the occupation probability of the conduction band at room temperature, which is

$$\simeq e^{\in g/2kT}$$

$$\simeq e^{-100} \approx 10^{-44}$$

that is, 1 atom in 10^{44} contributes an electron to the electrical conductivity.

Thus in a sample containing an order of 10^{20} atoms, there may be 10^{11} conduction electrons in a semiconductor, 10^{20} in a conductor (metal), and none in an insulator.

2. ***Find the temperature at which the number of electrons in the conduction band of a semiconductor increases by a factor of 10 over the number of electrons in it at room temperature. The band gap for the semiconductor is 0.67 eV.***

(At room temperature, $kT \sim \dfrac{1}{40}$ ***eV).***

Solution : The density of electrons in the conduction band at temperature T is given by

$$n_e = K_e\, e^{-\in_g/2kT},$$

where K_e is a constant and $\in g$ is the energy gap for the semiconductor.

At room temperature $kT \sim \dfrac{1}{40}$ eV = 0.025 eV; and $\in g$ = 0'67 eV (given). Thus, at room temperature, we have

$$ne = Ke\, e^{-0.67\ \text{eV}/2 \times 0.025\ \text{eV}} \qquad \ldots (i)$$

Let T be the temperature at which the density becomes $10\ n_e$.

Then

$$10\ n_e = K_e\ e^{-0.67\ \text{eV}/2kT\ (eV)} \qquad \ldots (ii)$$

Dividing eq. (*ii*) by (*i*), we get

$$10 = \frac{e^{-0.67\ eV/2kT\ (eV)}}{e^{-6\ eV/0.05\ eV}}$$

or $$\log e\ 10 = \frac{0.67\ \text{eV}}{0.05\ \text{eV}} - \frac{0.67\ \text{eV}}{2kT\ (\text{eV})}$$

or $$2.30 = 13.4 - \frac{0.67}{2kT}$$

or $$\frac{0.67}{2kT} = 13.4 - 2.30 = 11.1$$

or $$kT = \frac{0.67}{2 \times 11.1} = 0.030\ \text{eV}.$$

Now $$k = 1.38 \times 10^{-23}\ \text{joule}/K = \frac{1.38 \times 10^{-23}}{1.60 \times 10^{-19}}$$

$$= 0.86 \times 10^{-4}\ \text{eV/K}.$$

$\therefore$ $$T = \frac{0.030\ \text{eV}}{0.86 \times 10^{-4}\ \text{eV/K}} = 350\ \text{K}.$$

3. ***Find the concentration of germanium atoms in a pure Ge crystal from the following data : atomic weight of Ge = 72.6, density = $5.32 \times 10^3\ kg/m^3$ and Avogadro number $N_A = 6.02 \times 10^{23}$.***

Find also the concentration of Ge atoms relative to electron-hole pairs if the intrinsic carrier density at room temperature is $2.4 \times 10^{19}/m^3$.

Solution : By Avogadro's law, the number of Ge atoms per meter^3 of the crystal is given by

$$n_A = \frac{N_{A\rho}}{M},$$

where ρ is the density and M is atomic weight. Thus

$$n_A = \frac{6.02 \times 10^{23} \text{ atoms / gm} - \text{atom} \times 5.32 \text{ gm / cm}^2}{72.6 \text{ gm / gm - atom}}$$

$$= 4.4 \times 10^{22} \text{ atoms/cm}^3$$

$$= 4.4 \times 10^{28} \text{ atoms/m}^3.$$

The intrinsic concentration of electron-hole pairs at room temperature is $ni = 2.4 \times 10^{19}/\text{m}^3$.

$$\therefore \frac{n_A}{n_i} = \frac{4.4 \times 10^{28}}{2.4 \times 10^{19}} = 2 \times 10^9 \text{ germanium atoms per electron-}$$

hole pair.

4. ***The mobilities of electrons and holes in a sample of intrinsic germanium at room temperature are 0.36 and 0.17 m^2/V-s respectively. If the electron and hole densities are each equal to $2.5 \times 10^{19}/m^3$, calculate the electrical conductivity and resistivity of germanium.***

Solution : With two charge-carriers, the expression for the electrical conductivity is

$$\sigma = n_e \, e \, \mu_e + n_h \, e \, \mu_h \, ,$$

where n_e and n_h are the densities of electrons and holes and μ_e and μ_h are the corresponding mobilities.

In an intrinsic (pure) semiconductor, the number of electrons is just equal to the number of holes, that is,

$$n_e = n_h = ni \text{ (say)},$$

called 'intrinsic carrier density'.

Therefore, the (intrinsic) conductivity is

$$\sigma = ni \, e \, (\mu_e + \mu_h)$$

$$= (2.5 \times 10^{19}/\text{m}^3) \, (1.6 \times 10^{-19} \text{ C}) \, (0.36 + 0.17) \text{ m}^2/\text{V-s}$$

$$= 2.12 \frac{\text{coul / meter}}{\text{volt - sec}}$$

$$= 2.12/\text{ohm-meter.}$$

[$\because$ volt-sec = ohm-coul]

The resistivity is

$$\rho = \frac{1}{\sigma} = \frac{1}{2.12} = 0.47\ \Omega\text{-}m.$$

5. ***Calculate the intrinsic conductivity and resistivity of pure silicon at 300 K, assuming intrinsic carrier (electron-hole pair) density at this temperature to be 15 X $10^{16}/m^3$. The electron and hole mobilities in silicon are 0.135 and 0.048 m^2/volt-sec respectively.***

[**Ans.** 4:4 × $10^{-4}/\Omega$-m, 2.27 × 10^3 Ω-m.]

6. ***Find the resistance of an intrinsic germanium rod 1 cm X 1 mm x 1 mm at 300 K. For Ge : $n_i = 2.5 \times 10^{13}/cm^3$, $\mu_e = 3900$ cm^2/volt-sec and u_h = 1900 cm^2/volt-sec at 300 K.***

Solution : For an 'intrinsic' semiconductor, the expression for the conductivity is

$$\sigma = ni\ e(\mu_e + \mu_h)\ ,$$

where n_i is the intrinsic carrier density and μ_e and μ_h are electron and hole mobilities.

Given $ni = 2.5 \times 10^{13}/\text{cm}^3 = 2.5 \times 10^{19}/\text{m}^3$; $\mu_e = 3900\ \text{cm}^2$/volt-sec = 0.39m/volt-sec, μ_h = 0.19 m^2/volt-sec and e = 1.6 × 10^{-19} coulomb.

$$\therefore \sigma = (2.5 \times 10^{19}/\text{m}^3)\,(1.6 \times 10^{-19}\ \text{coul})\,(0.39 + 0.19)\text{m}^2/\text{volt-sec}$$

$$= 2.32/\text{ohm-'meter.}$$

The resistivity is

$$\sigma = \frac{1}{\sigma} = \frac{1}{2.32} = 0.43\ \text{ohm-meter.}$$

The length of the rod is l = 1 cm = 10^{-2} meter, and area of cross-section is A — 1 $(mm)^2$ = $10^{-6}m^6$. Its resistance is

$$R = \rho \frac{1}{A} = (0.43 \text{ ohm-m}) \frac{10^{-2} \text{ m}}{10^{-6} \text{ m}^2}$$

$$= 4.3 \times 10^3 \text{ ohm.}$$

7. ***A semiconductor has 6 x 10^{19} electrons and 7 x 10^{20} holes per meter^3. If the mobilities of electrons and holes are 0.10 m^2/V-s and 0.06 m^2/V-s respectively, calculate the conductivity of the semiconductor.Ans. 7.68 mho/m.***

8. ***Estimate the binding energy and the orbital radius of the donor (fifth) electron of arsenic in a germanium crystal. The dielectric constant of germanium is 16 and the effective mass of the electron is m = 0.2 m.***

Solution : When a pentavalent As atom replaces a Ge atom in the crystal lattice, four of the five valence electrons of the As atom form covalent bonds with the neighbouring Ge atoms while the "fifth" electron orbits around the positive core of the As^+ ion, behaving like the electron in the ground state of a hydrogen-like atom (Fig.). The chief difference is that this fifth electron moves in a lattice (not in vacuum), and so the dielectric constant K of Ge should be taken into account, and the mass of the electron m should be replaced by its effective mass m. Using hydrogen-like formula (taking n = 1) the binding energy of the electron is given by

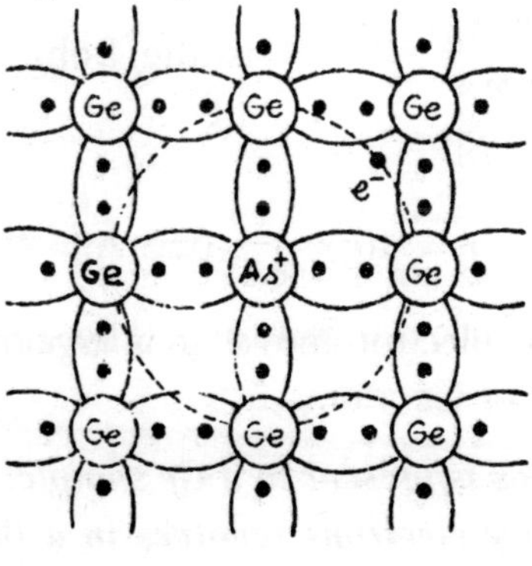

$$E = -\frac{1}{K^2}\frac{m^* e^4}{8\epsilon_0^2 h^2}$$

Substituting the values :

$$E = \frac{1}{(16)^2}\frac{(0.2 \times 9.1 \times 10^{-31}\ kg)(1.6 \times 10^{-19}\ coul)^4}{8\left(8.85 \times 10^{-12}\ \dfrac{coul^2}{nt\text{-}m^2}\right)(6.63 \times 10^{-34}\ joule\text{-}sec)^2}$$

$$= -\ 1.7 \times 10^{-21} \text{ joule}$$

$$= -\ \frac{1.7 \times 10^{-21}}{1.6 \times 10^{-19}} \simeq -\ 0.01 \text{ eV}.$$

The energy required for the fifth electron to leave the donor As atom and enter the conduction band is very small, only 0.01 eV. This is because of the high dielectric constant of Ge and a small effective mass of the electron in the crystal lattice.

The radius of the orbit is given by

$$r = K\frac{h^2 \epsilon_0}{\pi m^* e^2}$$

$$= K\frac{m}{m^*}a_0,$$

where a_0 $\left(= \dfrac{h^2 \epsilon_0}{\pi m e^2} = 0.53\ \text{Å}\right)$ is the Bohr smallest radius. Thus

$$r = 16 \times \frac{1}{0.2}\ 0.53\ \text{Å} = 42.4\ \text{Å}.$$

So the "fifth" electron moves in a *large* orbit that contains a large number of Ge atoms.

9. ***Phosphorus is present in a Ge sample. Assume that one of its five valence electrons revolves in a Bohr orbit around***

each P+ ion in the Ge lattice. If the effective mass of the electron is 0.17 m and the dielectric constant of Ge is 16, find the energy and the radius of the Bohr orbit of the electron. The energy gap between the valence and conduction bands in Ge is 0.67 eV. How does the ionisation energy of the above electron compare with this energy and with kT at room temperature?

Ans. $\simeq$ 0.01 eV, 50 Å. The ionisation energy is much smaller than the energy gap of 0.67 eV, but not very far from the 0025 eV value of kTat 27°C.

10. ***An intrinsic semiconductor has a carrier density of J0^{20} per m^3 at room temperature. It is doped with arsenic atoms so that there are 10^{21} arsenic atoms per in^3. Calculate the density of Conduction in hole carriers at room temperature, assuming that all the donor atoms[are | ionised.***

Solution : Arsenic is a donor impurity whose concentration is $N_d = 10^{21}/\text{m}^3$. The intrinsic carrier density is $ni = 10^{20}/\text{m}^3$.

In the n-type semiconductor, the hole concentration is given by

$$n_h \simeq \frac{n_i^2}{N_d} \simeq \frac{\left(10^{20}\right)^2}{10^{21}} \simeq 10^{19}/m^3.$$

11. ***An intrinsic sample of germanium has a hole density of 10^{19}/cm^3 at room temperature. When doped with antimony, the hole density decreased to 10^{11}/cm^3 at the same temperature. Calculate the density of the majority charge carriers.***

Solution : The *intrinsic* electron and hole densities are same:

$$ni = 10^{13}/\text{cm}^3.$$

Let n_e and n_h be the electron and hole densities in doped germanium. (Antimony is 'donor' which reduces the hole density). Then by mass-action law, we have

$$n_e \times n_h = ni^2$$

or $$n_e = \frac{N_i^2}{n_h} = \frac{\left(10^{13}\right)^2}{10^{11}} = 10^{15}/cm^3.$$

In n-type semiconductor electrons are the majority carriers.

12. ***The resistivity of pure silicon is 2300 Ω-m and the mobilities of electrons and holes in it are 0.135 and 0.048 m^2/V-s respectively. Find the electron and hole concentrations, and the resistivity of a specimen of silicon doped with 10^{19} atoms of phosphorus per meter^3.***

Solution : The intrinsic conductivity of a (pure) semiconductor is given by

$$\sigma i = ni\ e\ (\mu_e + \mu_h),$$

where n_i is the intrinsic carrier concentration, and μ_e, and μ_h are electron and hole mobilities. The corresponding resistivity is

$$\rho i = \frac{1}{\sigma_i} = \frac{1}{n_i\ e\left(\mu_e + \mu_h\right)}.$$

$$\therefore \qquad n_i = \frac{1}{\rho_i\ e\left(\mu_e + \mu^h\right)}$$

$$= \frac{1}{\left(2300\ \text{ohm} - \text{m}\right)\left(1.6\times 10^{-19}\ \text{coul}\right)\left(0.135 + 0.048\right) m^2\ /\ \text{volt-sec}}$$

$$= 1.48 \times 10^{16}/m^3.$$

The phosphorus is a "donor" with concentration, $N_d = 10^{19}/\text{m}^3$. In the doped n-type semiconductor, the free electron concentration is given by (assuming that all the donor atoms have been ionised)

$$n_e \simeq N_d \simeq 10^{19}/\text{m}^3,$$

and the hole concentration is

$$n_h \simeq \frac{n_i^2}{N_d} \simeq \frac{\left(1.48 \times 10^{16}\ /\ \text{m}^3\right)^2}{10^{19}\ /\ \text{m}^3}$$

$$\simeq 2.2 \times 10^{13}/\text{m}^3.$$

The conductivity of a doped semiconductor is given by

$$\sigma = (n_e \mu_e + n_h \mu_h)\, e.$$

In n-type semiconductor, electrons are the 'majority carriers' ($n_e >> n_h$) so that we can write

$$\sigma\ si\ n_e \mu_e e \simeq N_d \mu_e e$$

$$\simeq 10^{19}/\text{m}^3 \times 0.135\ \text{m}^2/\text{volt-sec} \times (1.6 \times 10^{-19}\ \text{coul})$$

$$\simeq 0.216/\text{ohm-m}.$$

The corresponding resistivity is

$$P = \frac{1}{\sigma} = \frac{1}{0.216/\text{ohm-m}} \simeq 4.63\ \text{ohm-m}.$$

The resistivity of the doped silicon is vastly different from the intrinsic resistivity.

13. ***A silicon wafer is doped with phosphorus of concentration 10^{13} atoms/cm^3. If all the donor atoms are active, what is the resistivity at room temperature? The electron mobility is 1350 cm^2/volt-sec. The charge on the electron is 1.6×10^{-19} coul.***

Hint : Convert in S.I. units.

[Ans. 4.63 Ω-m.]

14. ***Find the conductivity of p-type Ge crystal which is doped with acceptor atoms of concentration 2 X 10^{17} atoms/ cm^3 and all acceptor atoms are active. Given μ_h=1900 cm^2/ volt-sec and e =1.6×10^{-19} coul.***

Solution : In a p-type semiconductor, the hole concentration n_h is roughly equal to the acceptor concentration, that is

$$n_h \simeq N_a \simeq 2 \times 10^{17}/\text{cm}^3 \simeq \times 10^{23}/\text{m}^3.$$

The conductivity is given by

$$\sigma = (n_e \mu_e + n_h \mu_h) e.$$

In p-type semiconductor ($n_h >> n_e$) the holes are the 'majority carriers' and so we can write,

$$\sigma \simeq n_h \mu_h e$$

$$\simeq (2\times)0^{23}/m^3)\ (0.19\ m^2/\text{volt-sec})\ (1.6 \times 10\text{-}^{19}\ \text{coul})$$

$$\simeq 6100/\text{ohm-m}.$$

15. ***A Ge crystal is doped with 10" donor atoms/cm^3. Assuming that all donors are ionised, determine the resistivity of the doped sample. The electron mobility in Ge is 0.39 m^2/volt-sec and e = l.6 × 10^{-19} C.***

[**Ans.** 0.16 ohm-m.]

16. ***A germanium rod 10 mm long by 1 mm square in cross-section has been doped with a total of 5 × 10^{13} donor atoms at room temperature. Calculate the electron and hole densities and the conductivity. Also calculate the resistance between the square ends of the rod. The intrinsic carrier density in Ge is 2.4 × 10^{19}/m^3 and electron mobility is 0.39 m^2/V-sec.***

Solution : The length of the rod is $l = 10$ mm $= 10^{-2}$ m, area of cross-section is $A = 1\ mm^2 = 10^{-6}\ m^2$ and the volume is $V = 10^{-2}\ \text{m} \times 10^{-6}\ \text{m}^2 = 10^{-8}\ m^3$.

The donor concentration is

$$\text{Nd} = \frac{5 \times 10^{13}\ \text{atom}}{10^{-8}\ \text{m}^3} = 5 \times 10^{21}\ \text{atoms/m}^3$$

and the intrinsic concentration is given to be

$$n_i = 2.4 \times 10^{19}\ \text{atoms/m}^3.$$

In the n-type semiconductor, the electron density is given by

$$n_e \simeq N_d \simeq 5 \times 10^{21}/\text{m}^3,$$

and the hole density is given by

The conductivity of the doped specimen is

$$\sigma \simeq n_e \, e \, \mu_e + n_h \, e \, \mu_h$$

$$\simeq n_e \, e \, \mu_e \qquad [\because n_e >> n_h]$$

$$\simeq (5 \times 10^{21}/\text{m}^3)\,(1.6 \times 10^{-19}\text{ coul})\,(0.39\text{ m}^2/\text{V-sec})$$

$$\simeq 312/\text{ohm-m}.$$

The resistivity is

$$P = \frac{1}{\sigma} = \frac{1}{312} = 3.2 \times 10^{-2}\text{ ohm-m}.$$

The resistance is

$$R = \rho \frac{1}{A}$$

$$= (3.2 \times 10^{-3}\text{ ohm-m}) \frac{10^{-2}\text{ m}}{10^{-6}\text{ m}^2}\ 32\text{ ohm}.$$

14

Distribution of Energy

Fermi-Dirac Energy Distribution

In a metallic solid, the valence electrons are *loosely* bound to the individual atoms, and consequently travel *freely* throughout the volume of the metal. These are the 'conduction electrons'. We can treat these electrons as an ideal electron "gas" because their mutual repulsion is, on the average, cancelled by the attractions of the positive ions of the metal. The positive ions, however, create a potential which is same everywhere within the metal and zero outside. Therefore, the conduction electrons may be regarded to live within a "potential well" with the metal boundaries acting as high potential walls, as shown in Fig. below.

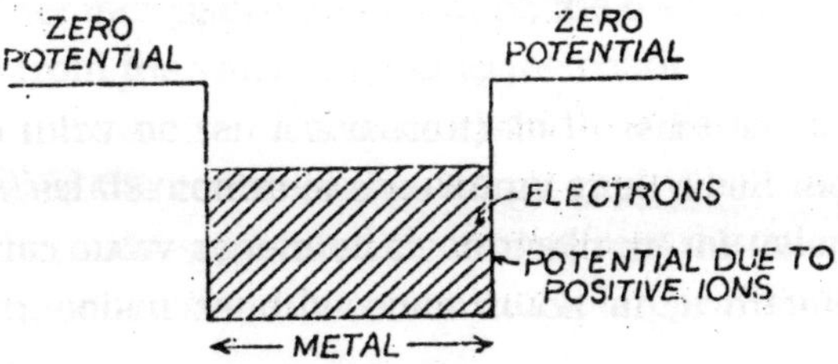

We have to find the distribution of electron energies in a metal. Since electrons are Fermi particles, they obey Pauli's exclusion principle and Fermi-Dirac distribution law.

According to this law, the number of electrons n_i with the energy $\in_i$ is given by

$$n_i = \frac{gi}{e^{\alpha}\, e^{\in i/kT} + 1}$$

where g_i is the number of quantum states having the same energy

In a solid, there are 10^{22} to 10^{23} atoms/cm^3 and the same number of valence electrons (if the atoms are monovalent). As such, there is an enormous number of quantum states which the electrons can occupy, that is, there is almost a continuous distribution of electron energies in the solid. In this situation the F-D distribution law becomes

$$n\,(\in)\,\mathrm{d}\in = \frac{g\,(\in)\,d\in}{e^{\alpha}\, e^{\in i/kT} + 1} \qquad (i)$$

where $n(\in)\,d\in$ is the number of electrons having energies between $\in$ and $\in + d\in$, and $g\,(\in)\,d\in$ is the number of quantum states available to these electrons.

Since the electrons are confined inside the metal, their wave properties will limit the energy values they may have. It can be shown that the number of allowed energy values between $\in$ and $\in + d\in$ is

$$\frac{4\sqrt{(2)}\,\pi\, V\, m^{3/2}}{h^3} \in^{1/2} d\in,$$

where m is the mass of electron and V is the volume of the electron gas. Since there are two possible spin states ($m_s = +\frac{1}{2}$ and $m_s = -\frac{1}{2}$) for an electron, each energy value can be had by two electrons, in accordance with exclusion principle.

Therefore, the number of energy states available to the electrons having energies between $\in$ and $\in + d\in$ is given by %

$$g(\in)\, d\in \;=\; \frac{8\sqrt{(2)}\,\pi V m^{3/2}}{h^3}\, \in^{1/2} d\in \qquad \text{... (ii)}$$

Substituting this value of $g(\in)\, d\in$ in eq. (*i*), we get

$$n(\in)\, d\varpi = \frac{8\sqrt{(2)}\,\pi V m^{3/2}}{h^3}\, \frac{\in^{1/2} d\in}{e^{\alpha} e^{\in/kT} + 1}.$$

In F-D distribution, the parameter a is strongly dependent on temperature *T*, and we write

$$\alpha = -\frac{\in_F}{kT}, \qquad \text{... (iii)}$$

where $\in_F$ is called the 'Fermi energy'. The last expression now becomes

$$\boxed{n(\in)\, d\in = \frac{8\sqrt{(2)}\,\pi V m^{3/2}}{h^3}\, \frac{\in^{1/2} d\in}{e^{(\in - \in_F)/kT} + 1}} \qquad \text{... (iv)}$$

This is the formula for the number of electrons having energies between $\in$ and $\in + d\in$ in a metal of volume *V* at Kelvin temperature *T*.

At the absolute zero ($T = 0$) there are two situations :

(i) For $\in < \in_F$, $e^{(\in - \in F)/kT} = e^{-\infty} = 0$,

$$\text{so that } n(\in)\, d\in = \frac{8\sqrt{(2)}\,\pi V m^{3/2}}{h^3}\, \in^{1/2} d\in,$$

which represents a parabolic curve.

(ii) For $\in > \in_F$, $e^{(\in - \in F)/kT} = e^{\infty} = \infty$, so that

$$n(\in)\, d\in = 0.$$

This means that no electrons have energy greater than $\in_F$ at the absolute zero. In other words, *the Fermi energy $\in_F$ is the* *maximum energy that a free electron in the metal can have at absolute zero.*

The plot of $n(\in)$ against $\in$ at $T = 0$ is shown in Fig. below. After rising along a parabola from the origin, it drops abruptly to zero at $\in = \in_F$. This distribution of the electron energies at $T = 0$ is understandable. The electrons occupy the lowest available energy states, but due to exclusion principle, each energy level can take only two electrons, so the electrons go to higher and higher energy levels until all are accommodated. The highest energy occupied is $\in_F$.

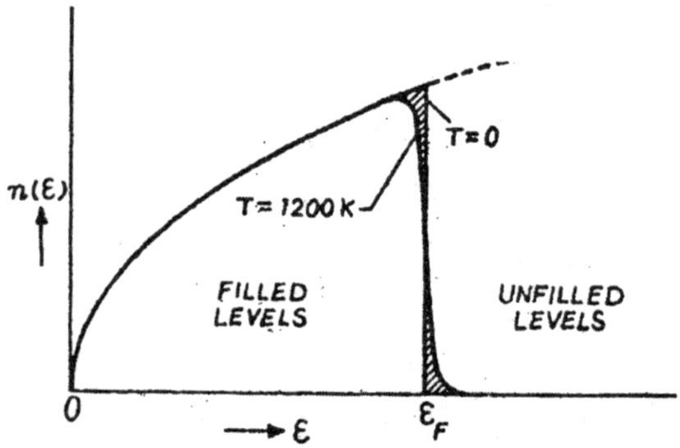

A typical plot of $n(\in)$ against $\in$ at a much higher temperature $T = 1200$ K is also shown in Fig. Since at any temperature the area under the corresponding curve gives the total number of electrons in the metal, the two shaded areas must be equal. It is seen that *even a considerable rise in temperature changes the energy state of only a very small fraction of the electrons.*

Calculation of Fermi Energy

Suppose a given metal contains N free electrons. We can calculate its Fermi energy $\in_F$ by filling up its energy states. At $T = 0$, starting from $\in = 0$, all quantum states upto $\in = \in_F$ are filled while all higher than $\in_F$ are vacant. The total number of free electrons is equal to the total number of quantum states upto the energy $\in_F$, because each state is limited to one electron (exclusion principle). That is,

$$N = \int_0^{\in_F} g(\in) d\in$$

Substituting the value of $g(\in) d\in$ from eq. (*ii*), we have

$$N = \frac{8\sqrt{(2)}\,\pi\, Vm^{3/2}}{h^3} \int_0^{\in_F} \in^{1/2} d\in$$

$$= \frac{16(2)\,\pi\, Vm^{3/2}}{3h^3} \in F^{3/2}$$

$$= \frac{8\pi V}{3}\left(\frac{2m}{h^2}\right)^{3/2} \in F^{3/2}$$

Thus
$$\boxed{\in F = \frac{h^2}{2m}\left(\frac{3N}{8\pi V}\right)^{2/3}}$$

This is the expression for the Fermi energy of a metal (strictly speaking at $T = 0$, but holding quite excellently upto $T \leq \in_F/k$). The quantity N/V is the density of free electrons. Thus the Fermi energy is independent of the size of the metal.

The Fermi energy ranges from 1.53 eV for cesium to 11.8 eV for aluminium.

The Fermi energy can be written in terms of Fermi temperature T_F which is defined by the relation

$$\in_F = k\, TF,$$

where k is Boltzmann's constant.

Electronic Specific Heat : Experimentally, the molar specific heat of metals at ordinary, and also at quite high, temperatures does not exceed a value of $3R$ (Dulong-Petit value). This specific heat is attributed to the vibration of the *atoms* in the metallic crystal lattice.

Now, in a typical metal, each atom contributes one electron to the common "electron gas", so that in one mole of the metal there are N free electrons, where N is Avogadro number. If these conduction electrons behave like the molecules of an ideal gas, each would have an average kinetic energy of (3/2) kT. Thus the (internal) electron energy per mole of the metal would be

$$Ee = \frac{3}{2} N kT = \frac{3}{2} RT. \qquad \left[\therefore k = \frac{R}{N}\right]$$

The molar specific heat of the electrons should therefore be

$$Cve = \left(\frac{\partial E_e}{\partial T}\right)_v = \frac{3}{2} R,$$

and the total specific heat of the metal should be

$$Cv = 3\text{R} + \frac{3}{2} R = \frac{9}{2} R,$$

and not merely $3R$, as shown by experiments. It means that the free (conduction) electrons do not contribute to the specific heat at ordinary and high temperatures.

The failure of the free electrons in a metal to contribute to its specific heat follows directly from the Fermi Dirac energy distribution among the electrons. It is seen from the distribution curve (Fig.) that when a metal is heated, *only* those electrons which are occupying the top energy levels (near $\in_F$) absorb additional energy and go to higher levels. The vast majority of the electrons in the lower levels cannot absorb more energy because the levels above them are already filled. The number of top electrons is, however, only a fraction $kT/\in_F$ of the total. At room temperature (T = 300 K), $kT \approx$ 0025 eV and $\in_F$ ranges from 1.5 to 12 eV for different metals. Thus less than 1% of the electrons absorb energy. Their contribution to the specific heat, which is $\frac{\pi^2}{2}\left(\frac{kT}{\in_F}\right)$ R, is of the order of 10^{-2} R. This is negligible compared with the atomic (lattice) contribution of $3R$.

It is only at temperatures of the order of 10^4 K (beyond the melting point of metals) that the electronic contribution is comparable with the lattice contribution.

At very low temperatures, however, the electronic contribution exceeds the lattice contribution because the electronic specific heat remains proportional to T, whereas the atomic specific heat becomes proportional to T^3 (Debye law). At temperatures near the absolute zero, the specific heat of metals is actually observed proportional to T, thus verifying the electronic contribution and hence Fermi-Dirac distribution.

Average Free-electron Energy at Absolute Zero :The number of electrons in an electron-gas having energies between $\in$ and $\in + d\in$ is given by

$$n(\in)\, D\in = \frac{2\sqrt{2}\,\pi V m^{3/2}}{h^3} \frac{\in^{1/2}\, d\in}{e^{(\in - \in_F)/kT} + 1}, \qquad \ldots (i)$$

where m is the mass of electron and is the volume of the electron-gas.

If N is the total number of electrons, then the Fermi energy $\in_F$ is given by

$$\in_F = \frac{h^2}{2m}\left(\frac{3N}{8\pi V}\right)^{2/3}$$

or $$(\in_F)^{3/2} = \frac{h^3}{2\sqrt{2}m^{3/2}} \frac{3N}{8\pi V}$$

or $$= \frac{8\sqrt{2}\,\pi V m^{3/2}}{h^3} = \frac{3N}{2}(\in_F)^{-3/2}$$

Making this substitution in eq. (*i*), we get

$$n(\in)\, d\in = \frac{3N}{2}(\in_F)^{-3/2} \frac{\in^{1/2}\, d\in}{e^{(\in - \in_F)/kT} + 1}$$

At $T = 0$ all of the electrons have energies less than or equal to $\in_F$ (*i.e.* $\leq \in_F$), so that at $T = 0$, we have

$$e(\in - \in_F)/kT = e^{-\infty} = 0.$$

Therefore, at absolute zero

$$n(\in)\, d\in = \frac{3N}{2}(\in_F)^{-3/2} \in^{1/2} de.$$

Now, let us first find the total energy E_0 at absolute zero, which is

$$\begin{aligned} E_0 &= \int_0^{\in_F} \in n(\in)\, d\in \\ &= \frac{3N}{2}(\in_F)^{-3/2} \int_0^{\in_F} \in^{3/2}\, d\in \\ &= \frac{3N}{2}(\in_F)^{-3/2}\left(\frac{2}{5}\in_F{}^{5/2}\right) \\ &= \frac{3}{5} N \in_F. \end{aligned}$$

The average energy $\overline{\in_0}$ is this total energy divided by the total number N of electrons. Thus.

$$\overline{\in_0} = \frac{E_0}{N} = \frac{3}{5}\in_F.$$

PROBLEMS

1. ***Compute the Fermi energy for sodium, assuming that each sodium atom contributes one free electron to the electron gas. The density of sodium is 0971 gm/cm^3 and its atomic weight is 23. (h = 6.63 × 10^{-34} joule-sec, m = 9.11 × 10^{-31}kg and Avogadro number is 6.02 × 10^{23} atoms/mole).***

Solution: Let V be the volume of a piece of sodium and N the number of free electrons in it. The free electron density, N/V, in sodium is equal to the number of sodium atoms per unit volume, because each sodium atom contributes one free electron to the gas. This can, in turn, be found from the density and atomic weight of sodium.

The mass of 1 mole (or gm-atom) of an element is M, where M is the molecular (or atomic) weight of the element. Its volume is M/p, where ρ is the density of the element. It contains N_A molecules (or atoms); where N_A is Avogadro's number. Therefore, the number of molecules (or atoms) per unit volume is

$$\frac{N_A}{M\rho} = N_A\rho/M.$$

The electron density in sodium is thus

$$\frac{N}{V} = \frac{N_A\rho}{M}$$

$$= \frac{(6.02 \times 10^{23} \text{ atoms / mole})(0.971 \text{ gm / cm}^3)}{23\text{gm / mole}}$$

$$= 2.54 \times 10^{22} \text{ atoms/cm}^3$$

$$= 2.54 \times 10^{28}\text{atoms/m}^3$$

$$= 2.54 \times 10^{28} \text{ electrons/m}^3.$$

The Fermi energy for sodium is

$$\epsilon_F = \frac{h^2}{2m}\left(\frac{3N}{8\pi V}\right)^{2/3}$$

$$= \frac{(6.63 \times 10^{-34} \text{ joule - sec})^3}{2(9.11 \times 10^{31} \text{ kg})}\left(\frac{3 \times 2.54 \times 10^{28} / m^3}{8 \times 3.14}\right)^{2/2}$$

$$= 5.0 \times 10^{-19} \text{ joule}$$

$$= \frac{5.0 \times 10^{-19}}{1.6 \times 10^{-19}} = 3.1 \text{ eV}.$$

2. ***There are 2 54 × 10^{22} free electrons per cm^3 in sodium. Calculate its Fermi energy, Fermi Texocity and Fermi temperature. (h = 6.63 × 10^{-34} joule-sec, m = 911 × 10^{-31} kg, k =1.38 × 10^{-23} joule/K, 1 eV=1.6 × 10^{-19} joule).***

Solution : The Fermi energy of a metal is given by

$$\in_F = \frac{h^2}{2m}\left(\frac{3N}{8\,\pi\,V}\right)^{2/3}$$

where m is the mass of electron and N is the number of free electrons in a volume V of the metal. Here

$$\frac{N}{V} = 2.54 \times 10^{22}\ \text{cm}^3 = 2.54 \times 10^{28}/\text{meter}^3.$$

Substituting this and the values of m and h, we get

$$\in_F = \frac{\left(6.63 \times 10^{-34}\ \text{joule-sec}\right)^2}{2 \times \left(9.11 \times 10^{-31}\ \text{kg}\right)}\left(\frac{3 \times 2.54 \times 10^{28}\ /\ \text{m}^3}{8 \times 3.14}\right)^{2/3}$$

$$= 5.0 \times 10^{-19}\ \text{joule} = 3.1\ \text{eV}.$$

This is the (maximum) kinetic energy of the free electron at absolute zero. If v_F be the Fermi velocity, then

$$\frac{1}{2}\, m\, v_F^2 = \in F = 5.0 \times 10^{-19}\ \text{joule}$$

$$\therefore \qquad v_F = \sqrt{\left(\frac{2 \times 5.0 \times 10^{-19}\ \text{joule}}{9.11 \times 10^{-31}\ \text{kg}}\right)}$$

$$= 1.05 \times 10^6\ \text{m/sec}.$$

The Fermi temperature T_F is defined by

$$\text{TF} = \frac{\in_F}{K} = \frac{5.0 \times 10^{-19}\ \text{joule}}{1.38 \times 10^{-23}\ \text{joule/K}}$$

$$= 3.6 \times 10^4\ \text{K}.$$

3. ***Consider silver in the metallic state, with one free (conduction) electron per atom. Calculate its Fermi energy. The density of silver is 10-5 gm/cm³ and its atomic weight is 108. The Avogadro's number Is 6.02 × 10²³ atoms/mole. Take values of h and m from the last problem.***

[**Ans.** 5.5 eV.]

4. ***The density of zinc ($Is^2\ 2s^2\ 2p^6\ 3s^2\ 3p^6\ 3d^{10}\ 4s^2$) is 713 gm/cm³ and its atomic weight is 654. Calculate its Fermi energy. The effective mass of a free electron in zinc crystal is 77 × 10⁻⁸¹ kg and the Avogadro's number is 6.02 × l0²³ atoms/ gm-atom.***

Solution : The electron configuration of the ground state of Zn shows that each atom has 2 valence electrons which it contributes to the electron gas.

The mass of 1 gm-atom of Zn is 65.4 gm, so that the volume

$$V = \frac{\text{mass}}{\text{density}} = \frac{65.4}{7.13} = 9.17 \text{ cm}^3.$$

The number of Zn atoms in this volume is 6.02×10^{23}. Therefore, the number of atoms per unit volume

$$= \frac{6.02 \times 10^{23}}{9.17} = 6.56 \times 10^{22}/\text{cm}^3.$$

Since each atom contributes 2 valence electrons, the number of free electrons per unit volume is

$$\frac{N}{V} = 2 \times 6.56 \times 10^{22}$$

$$= 13.1 \times 10^{22}/\text{cm}^3 = 13.1 \times 10^{28}/\text{meter}^3.$$

Now, the Fermi energy of the metal Zn is

$$\epsilon_F = \frac{h^2}{2m}\left(\frac{3N}{8\pi V}\right)^{2/3},$$

where m is the (effective) mass of the electron in zinc. Substituting the values, we get

$$\in_F = \frac{\left(6.62 \times 10^{-34} \text{ joule-sec}\right)^2}{2 \times \left(7.7 \times 10^{-31} \text{ kg}\right)} \left(\frac{3 \times 13.1 \times 10^{28} / m^3}{8 \times 3.14}\right)^{2/3}$$

$$= \frac{\left(6.62 \times 10^{-34}\right)^2}{2 \times 7.7 \times 10^{-31}} \ (6.25 \times 10^{18})$$

$$= 17.78 \times 10^{-19} \text{ joule}$$

$$= \frac{17.78 \times 10^{-19}}{1.6 \times 10^{-19}} = 11 \text{ eV.}$$

5. *The Fermi energy in silver is 5.51 eV.(a) What is the average energy of the free electrons in silver at 0 K ? (b) At what temperature a classical free particle (such as ideal gas molecule) will have this kinetic energy ?(k = 1.38 × 10^{-23} joule/K and 1 eV= 1.6 × 10^{-19} joule).*

Solution: (*a*) The average energy of an electron in electron-gas at 0 K is

$$\overline{\in_0} = \frac{3}{5} \in_F,$$

where $\in F$ is the Fermi energy. Now, $\in F$ = 5.51 eV. Therefore

$$\overline{\in_0} = \frac{3}{5} \times 5.51 = 3.306 \text{ eV.}$$

(*b*) The kinetic energy of a classical particle at Kelvin temperature *T* is $\frac{3}{2}$ *kT*. Then, if the particle has an energy 3.306 eV, then we have

$$\frac{3}{2} kT = 3.306 \text{ eV} = 3.306 \times (1.6 \times 10^{-19}) \text{ joule,}$$

where $k = 1.38 \times 10^{-23}$ joule/K. Therefore

$$T = \frac{2}{3} \times \frac{3.306 \times 1.6 \times 10^{-19} \text{ joule}}{1.38 \times 10^{-23} \text{ joule/K}}$$

$$= 2.55 \times 10^4 \text{ K.}$$

6. The Fermi energy for lithium is 4.72 eV at absolute zero. Calculate the number of conduction electrons per unit volume in lithium. ($h = 6.63 \times 10^{-34}$ J-s, $m = 9.11 \times 10^{-31}$ kg).

Solution: The Fermi energy is

$$\epsilon_F = \frac{h^2}{2m}\left(\frac{3N}{8\pi V}\right)^{2/3}$$

where N is the number of conduction electrons in a volume V of the lithium metal. The number of conduction electrons per unit volume is therefore

$$\frac{N}{V} = \left(\frac{2m}{h^2}\right)^{3/2} \epsilon_F^{3/2} \frac{8\pi}{3}.$$

Then, with $\epsilon_F = 4.72 \text{ eV} = 4.72 \times 1.6 \times 10^{-19}$ joule, we have

$$\frac{N}{V} = \left[\frac{2 \times 9.11 \times 10^{-19}}{\left(6.63 \times 10^{-34} \text{ joule - sec}\right)}\right]^{3/2}$$

$$= \frac{\left(4.72 \times 1.6 \times 10^{-19} \text{ joule}\right)^{3/2} \times 8 \times 3.14}{3}$$

$$= 4.64 \times 10^{28}/\text{m}^3.$$

7. For lithium, the Fermi energy is 4.70 eV and the density of electrons is $4.6 \times 10^{28}/m^3$. Find the electron density for a metal with Fermi energy 2.35 eV.

Solution: The electron density for a metal with Fermi energy ϵF is given by

$$\frac{N}{V} = \left(\frac{2m}{h^2}\right)^{3/2} \epsilon_F^{3/2} \frac{8\pi}{3}$$

or

$$\frac{N}{V} \propto \epsilon_F^{3/2}$$

The electron density for the metal is

$$\left(\frac{N}{V}\right)_{metal} = 4.6 \times 10^{28}/\text{m}^3 \left(\frac{2.35}{4.70}\right)^{3/2}$$

$$= 1.6 \times 10^{28}/m^3.$$

15

Energy Band Structure

Solids Band Theory

A solid contains an enormous number of atoms packed closely together. Each atom, when isolated, has a discrete set of electron energy levels *1s, 2s, 2p, 3s, 3p,......* If we imagine all the N (say) atoms of the solid to be isolated from one another, then they would have completely coinciding sets of energy levels. That is, each of the energy levels of this N-atom system would have an N-fold degeneracy. The electrons fill the energy levels in each atom independently. As the atoms approach one another to form the solid, a continuously increasing interaction occurs between them which causes each of the levels to -'split" into N distinct levels. In practice, however, N is very large ($\approx 10^{23}/cm^3$). Therefore, the splitted energy levels become so numerous and so close together that they form an almost continuous "energy *band*".

The amount of splitting is different for different energy levels. In general, the lower levels are splitted less than the higher levels, the lowest levels remaining almost unsplitted.

The reason is that the electrons in lower levels are the "inner" electrons of the atoms, which are not significantly influenced by the presence of nearby atoms. On the other hand, the electrons in higher levels are the "valence" electrons whose wave functions overlap appreciably.

Figure below shows the formation of energy levels for some of the higher energy levels of isolated sodium atoms (whose ground-state configuration is $Is^2\ 2s^2\ 2p^6\ 3s^1$) as their interatomic distance decreases. (The dashed line indicates the observed interatomic separation in solid sodium). The 3s level is the first "occupied" level to be splitted into a band ; the 2p level does not begin to split until the interatomic distance becomes smaller than actually found in the solid sodium. (The levels Is and 2s do not split at all).

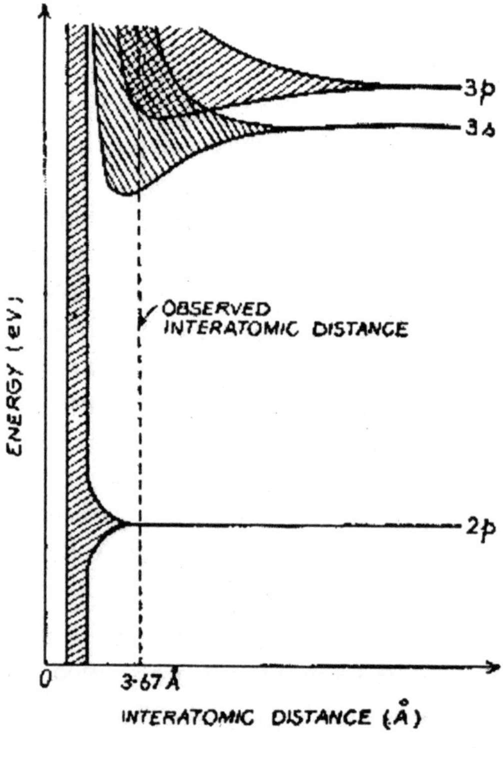

Now, the energy bands in a solid correspond to the energy levels in an atom. An electron in a solid can have only energies that fall within these energy bands. The various energy bands in a solid may or may not overlap depending upon the structure of the solid. If they do not overlap (Fig. a) then the intervals between them represent energies which the electrons in the solid cannot have. These intervals are called "forbidden bands" or "energy gaps". If, however, the adjacent energy bands in a solid overlap (Fig. *b*), the electrons have a continuous distribution of allowed energies.

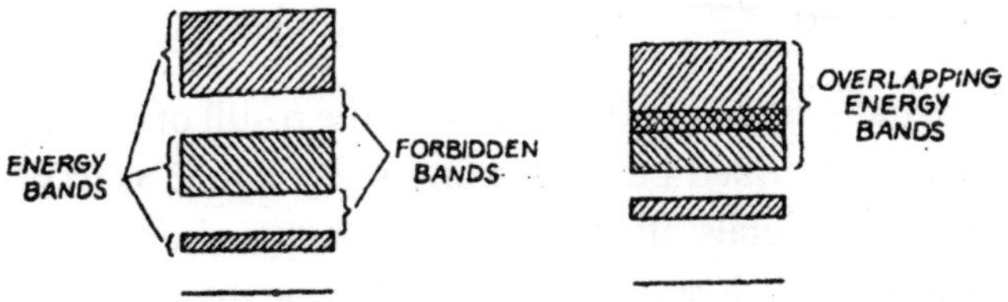

(a) (b)

***Solids Classification* ON *the Basis of Band Structure* :** The electrical properties of a solid depend upon its energy-band structure and the way in which the energy-bands are occupied by the electrons. In general, each energy band has a total of *N* individual levels, and each level can hold 2 × (2*l*+1) electrons so that the capacity of each band is *2(214- l) N* electrons. Thus the *\s, 2s, 2p, 3s*.......bands can hold *2N, 2N, 6N, 2N,* ...... .electrons respectively. Depending on the nature of band occupation by electrons and on the width of forbidden bands, all solids can be classified as conductors, insulators and semiconductors

***Conductors* :** In some solids there is a partially-fitted band above the completely-filled lower bands. Such a band is formed from partially-filled atomic levels as in case of alkali metals like sodium (Fig. a). A sodium atom has a single valence electron in its outer *3s* level. Therefore, of the *N* atoms in a solid piece of sodium, each contributes only one *3s* electron to the solid, and so there are only *N* (valence) electrons in the *3s* band. The valence band *3s* is thus only half full.

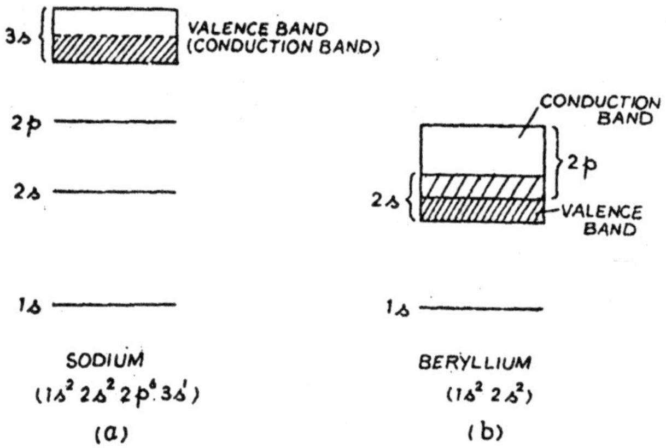

A partially-filled band may also be the result of overlapping of a completely-filled band and an empty band, as in case of alkaline-earth metals. In Fig. *b* are shown the energy bands of beryllium in which there is an overlap of the lower energy levels of the empty *2p* band with the upper energy levels of the completed 2s band. Those electrons which would occupy the highest energy levels in the *2s* band will actually go into the lowest levels of the overlapping *2p* band. Thus levels at the top of the *2s* band become unoccupied and the band is only partially-filled.

Now, suppose an electric field is applied across a piece of solid sodium (or beryllium). Then electrons in the partially-filled valence band easily acquire additional energy to move to the higher unoccupied energy levels *within the same band,* without crossing any energy gap. The additional energy is in the form of kinetic energy, and the moving electrons constitute an electric current. Sodium (or beryllium) metal is therefore a good conductor of electricity. Thus, *a partially-filled valence energy-band is a feature of conductors.*

An empty band into which electrons can pass is termed as 'conduction band'. In conductors, the valence band itself is the conduction band.

Insulators : In some solids, the valence band (containing the outer electrons of the atoms) is completely filled, while the next higher band separated by an energy gap of a few electron-volts is completely empty. Such a solid is an "insulator". Diamond and sodium chloride are typical examples of insulators.

Figure below shows the energy bands of diamond. There is an energy band completely filled with electrons (the valence band), and above it is an empty band (the conduction band) separated by a gap of 7eV. (The bands below the valence band are also completely filled).

At least 7 eV of energy must be provided to an electron in the diamond crystal in order to enter the conduction band where it can move freely. With kT = 0.025 eV at room temperature, valence electrons do not have enough thermal energy to cross the 7-eV gap.

DIAMOND
($1s^2 2s^2 2p^2$)

Now, if an electric field be applied, the electrons in the valence band would not accept energy to move within the band because there are no unoccupied levels in this band. They can, however, move to the higher empty band provided they get energy of about 7 eV to cross the gap. Since the electric field *cannot* give this amount of energy, the electrons do not acquire a directional motion. Diamond is, therefore, an insulator.

Semiconductors

Certain solids have the basic crystal structure of an insulator, but with a much smaller energy gap (of the order of an electron-volt) between the valence band and the conduction band. Such solids are known as semiconductors. Silicon and germanium, having energy gaps of 1.1 eV and 0.7 eV respectively, are typical examples of semiconductors.

The below given figure shows the energy bands of silicon. At room temperature, a few of its electrons in the valence band have sufficient kinetic energy of thermal motion to cross the narrow energy gap (forbidden band) and enter the conduction band above it. Hence, when an electric field is applied, the few electrons present in the conduction band acquire additional energy to move to the unoccupied levels within the same band. Similarly, a few of the many electrons present in the valence band move to the few available unoccupied levels in the same band. Hence there is a *limited* flow of current across the crystal. Thus, silicon has an electrical conductivity intermediate between those of conductors and insulators, and is therefore a semiconductor.

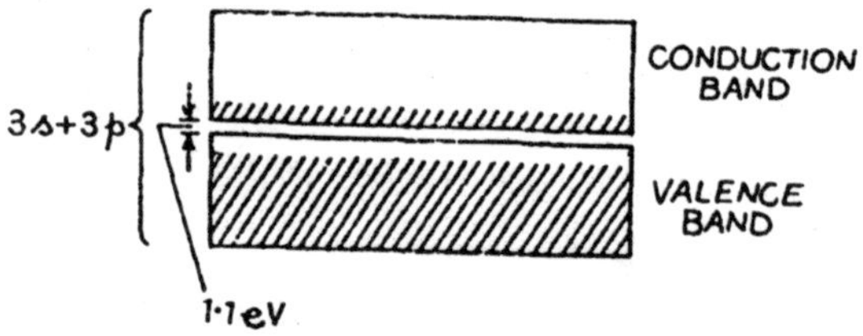

SILICON

$(1s^2 2s^2 2p^6 3s^2 3p^2)$

Relative Conductivities of Ge, Si and Diamond : At absolute zero, the semiconductors (Ge and Si) have the same energy-band structure as the insulators (diamond), that is, a *filled* valence band and above it an *empty* conduction band with an energy gap in between. Hence, at temperature near absolute zero, all the three Ge, Si and diamond are insulators.

As the temperature increases, some of the valence electrons gain enough thermal energy to jump over the energy gap and enter the conduction band. Thus electrons become available in the conduction band, and vacancies (or holes) in the valence band. They both cause conduction when electric field is applied. The probability of this happening increases with temperature, and it depends strongly on the width of the energy gap (forbidden band). In Ge and Si, the energy gaps are narrower, 07 and 1.1 eV, and so they become reasonably conducting at room temperature ; Ge being a little better than Si as conductor. On the other hand, the gap between the filled and the empty band in diamond is quite large, 7 eV. Hence diamond remains an insulator even at high temperatures.

Optical Properties of Solids : The optical properties of solids are closely connected with their energy-band structures. For example; conductors, semiconductors and insulators behave differently to light. This is because of the basic difference in their energy-band structures.

(a) *All Metals are Opaque to Light of all Wavelengths :* In metals (conductors), the valence energy band is either partially filled or (if completely filled) overlapped by an empty conduction band. Therefore, the valence electrons can acquire additional energy, *however small,* to move to the unoccupied energy levels of the *same* band. Hence, when light of *any* wavelength falls on a metal, the electrons absorb the light photons (energy) and are excited to the higher (unoccupied) energy levels. Thus light is absorbed, not transmitted through the metal. That is, the metal is opaque to light of all wavelengths. The characteristic luster of metals is due to there-radiation of light absorbed by their valence (free) electrons. If the metal surface is smooth the re-radiated light appears as a reflection of the incident light.

(b) *Semiconductors are Opaque to Visible Light but Transparent to Infrared Light :* In a semiconductor the valence band is full above which there is an empty conduction band with a small energy gap, of the order of 1 eV in between. Photons of visible light have energies roughly between 1 and eV, and so they are absorbed by the valence electrons which are excited to the conduction band. Hence the semiconductor is opaque to visible light. The photons of infrared light are, however, of much smaller energies and fail to excite the electrons in the valence band. Hence the infrared light passes through the semiconductor *i.e.* the semiconductor is transparent to infrared light.

(c) *Insulators are Transparent to Visible Light :* The electrons in the valence band of an insulator need more than 3 eV of energy to jump across the forbidden band to the conduction band. Diamond, for example, requires 7 eV of energy. Insulators therefore cannot absorb photons of visible light (having 1-3 eV energies) which passes through them. Many insulators are, however, opaque to ultraviolet light whose high-energy photons make electrons to cross the forbidden band.

Electrical Conductivity of Metals

The mechanism how a metal conducts electricity can be understood by the free-electron model. The crystal lattice of a metal consists of positive ions at the lattice points, and the valence electrons are *free* to move inside the crystal. In the absence of an applied electric field, the free electrons move in random directions. This is because the electrons frequently collide with the imperfections in the crystal lattice, which arise from the thermal vibrations of the ions about their equilibrium positions in the lattice and also from the presence of impurity ions. After each collision, the electrons are scattered in new directions with new speeds and this makes their motion random. If τ be the average time between the collisions of an electron, then

$$\tau = \frac{\lambda}{v_F},$$

where λ is the average distance between collisions (mean free path), and v_F is the speed of those electrons whose kinetic energy is equal to the Fermi energy. (Only electrons near the Fermi level contribute to the conductivity).

When an electric field is applied to a metal, the electrons modify their random motion in such a way that, on the average, they drift slowly in the direction *opposite* to that of the field (because their charge is negative) with a small speed vd. Let us calculate it.

The electric field E applied to an electron in the metal exerts on it a force eE which gives it an acceleration a, given by

$$a = \frac{eE}{m},$$

where e is the charge and m is the mass of electron. Let us consider an electron that has just collided with a lattice imperfection. Its drift speed has momentarily become zero and it would now move in a purely random direction, gaining a drift speed $a\ \tau$ just before its next collision. Its average drift speed during the interval τ is, therefore

$$vd = \frac{a\tau}{2} = \frac{aE\tau}{2m}.$$

Substituting the value of τ from above, we get

$$vd = \frac{eE\lambda}{2m\, v_F}$$

If n be the number of electrons per unit volume in the conduction band of the metal, then the current density./ (flow of charge per unit time across unit area) is

$$j = nev_d = \frac{ne^2E\lambda}{2mv_F}$$

The "resistivity" ρ of a metal is defined as

$$\rho = \frac{E}{j}$$

Therefore, we have

$$\boxed{\rho = \frac{2mv_F}{ne^2\lambda}.} \quad \text{... } (i)$$

This equation can be taken as a statement of Ohm's law, because the quantities v_F and λ, which determine ρ, do not depend on the applied field. Hence the resistivity ρ is a constant for a given metal at a given temperature.

If σ be the conductivity of the metal, then

$$\boxed{\sigma = \frac{1}{\rho} = \frac{ne^2\lambda}{2mv_F}.} \quad \text{... } (ii)$$

Let us define a measurable quantity "mobility" μ. It *is the drift velocity per unit electric field,* that is,

$$\mu = \frac{vd}{E} = \frac{e\lambda}{2mv_F}. \quad \text{... } (iii)$$

Making this substitution in eq. (*ii*), we get

$$\boxed{\sigma = ne\mu.} \quad \text{... } (iv)$$

In divalent metals as beryllium, zinc and cadmium; electrical conduction occurs by positive charge carriers as well. This arises due to transition of electrons from the filled valence band to the conduction band leaving vacancies, called *holes,* in the valence band. Each of these holes corresponds to the absence of an electron and behaves like a positive charge. As these vacancies are filled up by electrons moving under the influence of an electric field, new holes are created. That is, the holes move in a direction opposite to the electrons as if positive charge carriers were moving in the direction of the field. In fact, in these metals the mobility of holes (positive carriers) in the valence band is much greater than the mobility of electrons (negative carriers) in the conduction band; that is, the contribution to electrical conduction by positive carriers exceeds that by negative carriers.

If conduction occurs by electrons as well as by holes, then the conductivity is given by

$$\sigma = n_c e\mu_c + n_h e \mu_h,$$

where μ_e and μ_h are the mobilities of electrons and holes respectively, and n_e and n_h are their numbers per unit volume.

Conductivity Variations in Metals

When the temperature of a metal is raised above the room temperature, its conductivity decreases or the resistivity increases. This can be explained in the following way :

Metals have electrical resistivity due to the scattering of free electrons by the imperfections in the metallic lattice. Imperfections are of two kinds : (i) those arising from the thermal vibrations of the positive ions about their equilibrium positions in the lattice and (ii) those arising from structural defects such as presence of impurity ions. Therefore, the resistivity ρ of a metal can be written as

$$\rho = \rho_t + \rho_i$$

where ρt, is the resistivity caused by thermal imperfections and ρi that caused by impurity or structural imperfections. Of these, *pi* depends upon the temperature. As the temperature rises, the amplitude of ion vibrations increases, with the result that the scattering cross-section of the ions increases. Therefore, the resistivity ρ_t increases. The variation of ρ_i with temperature is of the form $\rho, \propto T^5$ at low temperatures and $\rho_t \propto T$ at high temperatures. ρ_t dominates except at very low temperatures.

The increase of metallic resistivity with temperature can be seen from the resistivity expression also, according to which

$$\rho \propto \frac{v_F}{\lambda}.$$

As temperature rises, the electron speed v_F increases and the mean free path A decreases (because the increased amplitude of vibration of the positive ions further obstruct the path of free electrons). Hence the resistivity ρ increases.

The resistivity ρ_i, on the other hand, is independent of temperature. It, however, dominates only in the temperature range about 0 - 6 K, because ρ, vanishes as $T \to 0$.

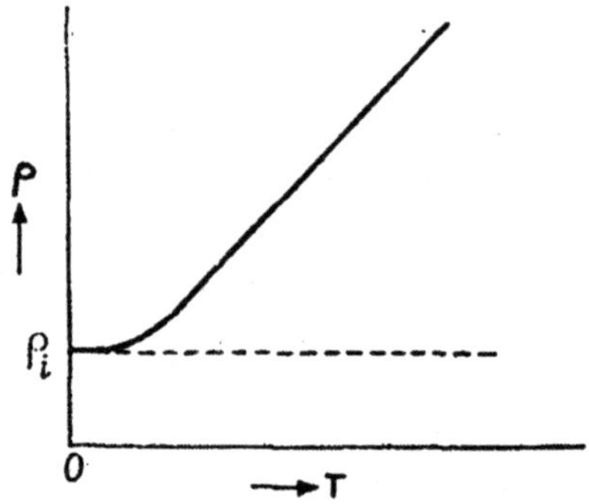

The variation of resistivity ρ with temperature T is shown in figure above.

Superconductivity

In 1911, Kammerlingh Onnes found that the electrical resistance of some metals, alloys and compounds drops suddenly to zero when the specimen is cooled below a certain temperature. This phenomenon is called "superconductivity and the cooled specimen is said to be a "superconductor",

The "critical" temperature T_c below which a material undergoes a transition from a state of normal conductivity to a superconducting state, is different for different materials. It varies from 23.kg for the alloy Nb_3 Ge to 0.01K for some semiconductors. Not all materials, however, superconduct. The normally good conductors like Cu, Ag, Au, Li, Na K, etc. do not show superconductivity even at temperature as low as a small fraction of 1K. (They are known as normal metals) below given the figure shows the resistivity at very low temperatures for a superconductor, tin (T_c = 3.7K); and a normal metal, silver. (Normal metals have a residual resistivity even near the absolute zero due to the structural imperfections in the crystal lattice.)

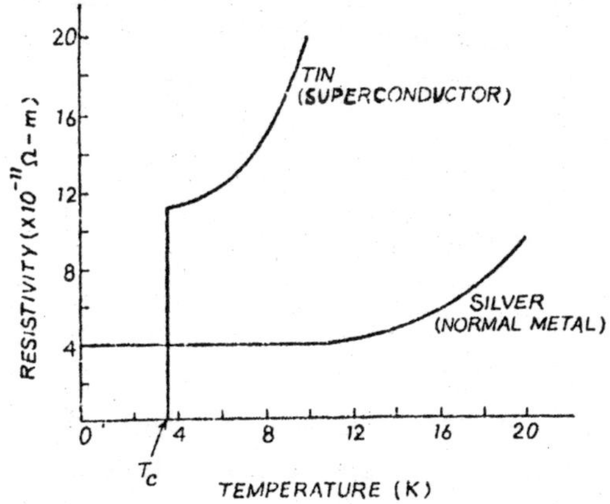

Electrical and Magnetic Behaviour of Superconductors : The superconducting state of a material is characterised by *zero* electrical resistance. Hence a superconductor can conduct electric current even in the absence of an applied voltage, and the current can persist for years without any detectable decay.

The magnetic properties of superconductors are equally dramatic. A bulk superconductor in a weak magnetic field acts as a perfect diamante, with zero magnetic induction in the interior. If a superconducting material is placed in a magnetic field (Fig. a) and is then cooled to below its critical temperature, it expels all the originally present magnetic flux from its interior (Fig. b). This is called 'Meissner effect'.

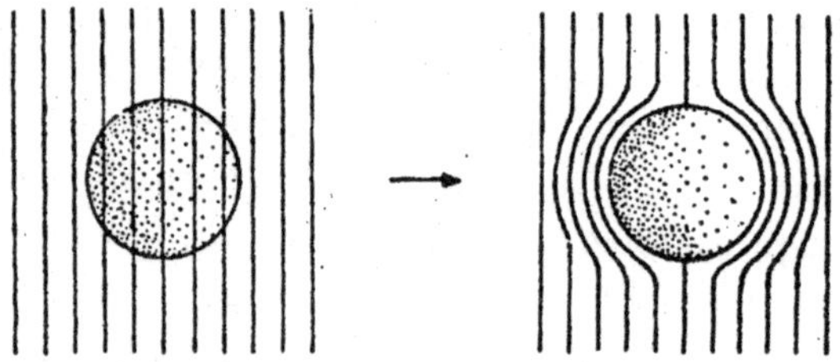

(a) (b)

Further, it is found that if the external field is increased beyond a certain value, called the 'critical field', the material ceases to be superconducting and becomes normal. The value of this critical field for a given material depends on the temperature.

Quantum Theory of Superconductivity **:** In 1957, Bardeen, Cooper and Schrieffer proposed a theory of superconductivity in which they included the electron-phonon and electron-electron interactions. We consider only a qualitative picture of this so-called BCS theory.

An electron in a solid passing by adjacent (positive) ions in the lattice exerts upon them Coulomb attractions which give them momentum and cause them to move slightly together. This region of increased positive-charge density propagates through the lattice as an elastic wave (phonon) carrying momentum supplied by the electron. If a second electron is passing by the moving region of increased positive-charge density, it will experience an attractive Coulomb interaction and absorb *all* the momentum carried by the moving region.

The net effect is that the two electrons have exchanged some momentum with each other. That is, they have interacted with each other through the lattice deformation (or phonon). This electron-electron interaction is attractive, because both the electrons participated in Coulomb attraction with the lattice. If the temperature is low enough and the two electrons have "anti-parallel" spins, the attraction between the electrons exceeds slightly the direct Coulomb repulsion between them.

Then the electrons are weakly bound together and form a so-called "Cooper pair". A large number of such pairs are formed at low temperature if there are enough conduction electrons lying just below the Fermi energy level.

When an external electric field is applied, the Cooper pairs move through the lattice, maintaining their order, as if the motion of each pair is locked into the motion of all the rest. Thus electrons move in highly correlated pairs, *without involving in the random scatterings from lattice imperfections* that cause electrical resistance. Hence the solid is a superconductor.

Ordinary metals can be good conductors when their conduction electrons have a weak interaction with the lattice. Superconductors, on the other hand, have a relatively strong electron-lattice-electron interaction. This is why the metals which are normally good conductors do not become superconductors even at extremely low temperatures.

Explanation of the Characteristics of Superconductors : The behaviour of a superconductor as a perfect diamagnet in an external magnetic field can now be explained. According to Lenz's law, when the magnetic flux through a circuit is changing, an induced current is established in such a direction as to oppose the change in flux. The external magnetic field does not penetrate the interior of the superconductor because in it the conduction electrons, moving in Copper pairs *without any resistance adjust* their motion to produce a counteracting magnetic field. Obviously, it is necessary for a superconductor to have a persisting (resistance less) current to keep the magnetic flux out when there is an external magnetic field. Thus, the two main characteristics of superconductors, namely the exclusion of magnetic flux and the absence of resistance to current flow, are related to each other.

Again, when the external magnetic field is increased beyond a certain critical value, the superconductor becomes normal (although it is at a temperature below T_c). This is also understandable. When the magnetic field is turned on, the superconductor acts to exclude this field from its interior. The energy decrease of the magnetic field appears as increased energy of the electrons in Cooper pairs. As the magnetic field is increased; the energy acquired by the superconductor also increases. Beyond the critical value of the field, the energy of the superconducting state becomes higher than the energy of the normal state, and the material becomes normal.

Applications : There are many applications of superconductivity.

(i) One is in the construction of superconducting electromagnets that carry large resistance less currents and therefore produce large magnetic fields ($\approx 5 : 10$ Tesla). If we use superconducting wire for magnet windings, currents of the order of 100 amperes can be carried by very fine wires. Thus small-size magnets can be constructed. The principle is illustrated in Fig. below, in which a superconducting coil is immersed in liquid helium (4 K). Starting from zero, the current is increased until the magnetic field B reaches the desired value. At that point the switch *S* is closed. The current now flows through the switch. The power supply is then turned off. The current will continue to flow through the coil indefinitely without resistive losses.

(ii) Superconducting cables can be used to transmit electric power over long distances without resistive losses. This would be economical, however, if the cost of keeping the cable below its critical temperature T_c is less than the value of the power that would be lost using ordinary cables. Materials are now available having $T_c \approx 20$ K.

(iii) Because superconductors are diamagnetic, they can be used to shield out unwanted magnetic flux, as in shaping the magnetic lens system of an electron microscope.

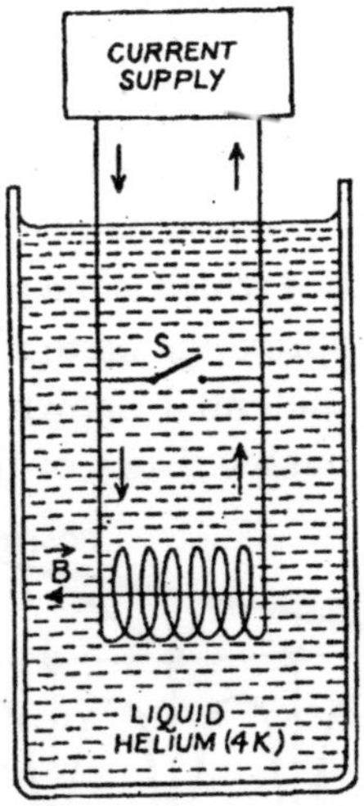

Motion of Electrons in a Periodic Lattice—Origin of Energy Bands and Forbidden Bands : The free-electron model of metals assumes the conduction electrons to move freely in a region of constant potential without interacting with the crystal Lattics. Although this model explains certain properties of metals, such a conductivity, specific heat, para-magnetism, etc., but it fails explain satisfactorily the properties of solids in general. Hence it needs to be modified.

In general, an electron in a solid moves in a region of *periodically-varying potential* (with the periodicity of the lattice) caused by the ion-cores situated at the lattice points, plus the average effect of all the other free electrons. This results in the diffraction of electrons by the lattice. When the de Broglie wavelength of the electron ($= h/p$) corresponds to a periodicity in the spacing of the ions, the electron interacts strongly with the lattice and goes Bragg "reflection". This limits the electron to certain ranges of momenta and, correspondingly, to certain ranges of energy (energy bands).

In order to find the allowed energies of electrons in solids we must solve the Schroedinger equation for an electron in a crystal lattice. The upper part of the below given figure shows the actual potential as seen by an electron in the crystal lattice in one dimension. Kronig and Penney suggested a simplified mode potential consisting of an infinite row of rectangular potential wells separated by barriers of width b, with space periodicity a which is the periodicity of lattice. Each well represents an approximation to the potential produced by one ion. This model potential, shown in the lower part of the figure, has all the characteristics of the actual potential.

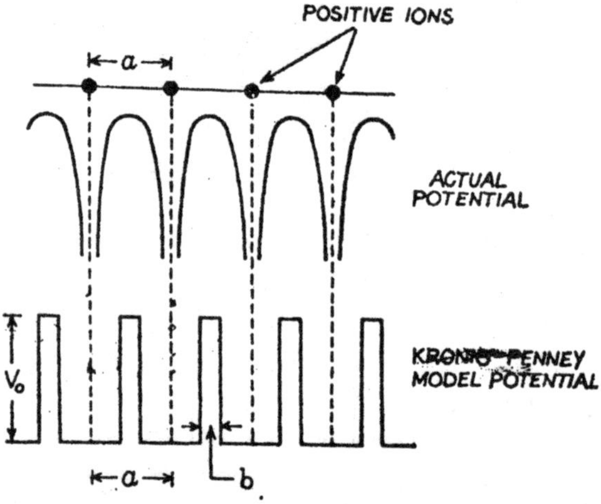

The solution to the Schroedinger equation for an electron in a periodic lattice can be found by Bloch's theorem. According to this theorem, the eigenfunction of the free-electron traveling wave, $\psi = e^{ikx}$ where $k\ (= 2\pi/\lambda)$ is propagation constant, is modified by the periodic potential to be of the form

$$\psi(x) = u_k(x)\, e^{ikx}$$

where the modulating factor $u_k(x)$is periodic with the same periodicity a as of the potential, that is,

$$u_k(x) = u_k(x + a) = u_k(x + na),$$

n being an integer. Hence, the effect of the periodicity is to modulate periodically the free-electron wave amplitude. The wave-function is

$$\psi(x, t) = uk(x)\, e^{i(kx-\omega t)}$$

where the second (exponential) factor describes a wave of wave length $\lambda = 2\pi/k$ which travels toward $+x$ if k is positive, and toward— x if k is negative. The exact form of the modulating factor depends upon the particular potential and the value of k.

On solving the Schroedinger equation of the electron for the Kronig-Penney potential under the condition that ψ and $d\psi/dx$ must be continuous at the boundaries of the well, a complicated expression for the allowed energies in terms of k of the electron is obtained which shows that gaps in energy occur at values given by

$$k = \pm \frac{\pi}{a}, \pm\frac{2\pi}{a}, \pm\frac{3\pi}{a}, \dots \dots$$

in which a is the space periodicity of the lattice. The figure below shows a relationship between energy $\in$ and wave number k for a one dimensional lattice. The dashed curve is the free-electron parabola. At the above values of k we get energy gaps, whereas for values of k not near these values the energies are much like that of a free electron. The origin of the allowed energy bands and forbidden gaps is apparent from the figure.

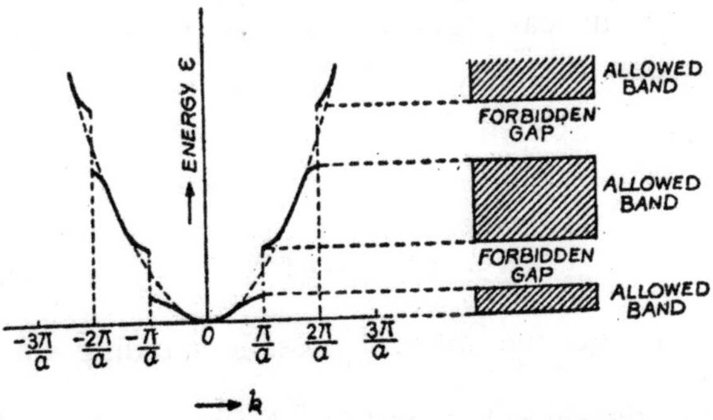

The occurrence of gaps can be understood in terms of "Bragg reflection" of the de Broglie waves associated with the electron moving through the lattice. The Bragg's condition for the reflection of waves is

$$2a \sin \theta = n\lambda, \qquad n = 1, 2, 3,......$$

where a is the spacing between the ions of the lattice. Since we are considering the lattice in one dimension (along the x-axis) only, the above equation becomes

$$2a = n\lambda.$$

But
$$\lambda = \frac{2\pi}{k}.$$

$\therefore$
$$2a = n\frac{2\pi}{k}$$

or
$$k = n\frac{\pi}{a} = \pm\frac{\pi}{a}, \pm\frac{2\pi}{a}, \pm\frac{3\pi}{a}, \ldots \ldots$$

where we have inserted ± signs because the incident wave can move along + x as well as along —x. These are just the values of k at which the gaps in the $\in$-k curve occur. The waves corresponding to values of k *not* satisfying the above condition travel almost freely. Those satisfying the condition however, are reflected resulting in "standing" waves. Let us consider an incident wave of unit amplitude corresponding to $k = \frac{\pi}{a}$. The eigenfunctions describing the incident and the corresponding reflected waves are e^{ikx} or $e^{i(\pi/a)x}$ and $e^{-(\pi/a)x}$ The two eigenfunctions can combine in two ways to give total eigenfunction, namely,

$$\psi_1 = e^{i(\pi/a)x} + e^{-i(\pi/a)x} \qquad \propto \cos\frac{\pi}{a}x$$

and
$$\psi_2 = e^{i(\pi/a)x} - e^{-i(\pi/a)x} \qquad \propto \sin\frac{\pi}{a}x.$$

Thus we get two *different* possible standing waves. The probability densities $|\psi_1|^2$ and $|\psi_2|^2$ for these waves are plotted in figure below, in which the origin (x = 0) has been taken at one of the ions. Evidently, $|\psi_1|^2$ is maximum at the positive ions ($x = 0, \pm a, \pm 2a$,......) and minimum in between. Reverse is for $|\psi_2|^2$. The potential energy of an electron in a lattice of positive ions is greatest midway *between* the ions and least *at* the ions. So an electron for $k = \frac{\pi}{a}$ can have two *different* energies E_1 and E_2 *a* associated with the standing waves ψ_1 and ψ_2. No other solutions are possible when $k = \frac{\pi}{a}$, and accordingly no electron can have an energy between E_1 and E_2. The difference $E_1 \sim E_2$ is the energy gap. This is the origin of all the energy gaps.

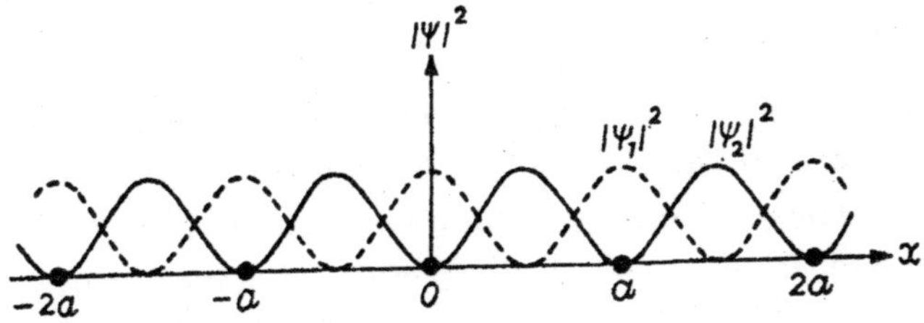

Looking again at the ∈-k curve, the range of k values between —π/a and + π/a defines the first Brillouin zone, those of k values between — $2\pi/a$ and — π/α and between $+\pi/\alpha$ and $+2\pi/a$ define the second Brillouin zone, etc. The first and second Brillouin zones of a two-dimensional square lattice are shown in figure below.

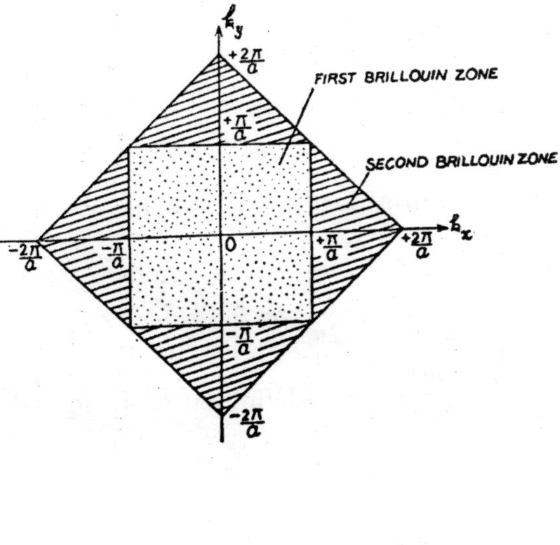

Effective Mass of Electron

An electron in a crystal interacts with the crystal lattice. Therefore, its behaviour towards external forces is different from that of a free electron. The deviation of the electron behaviour in the crystal lattice from the free-electron behaviour can be taken into account simply by considering the electron to have an "effective mass" *m*, rather than its free-space mass *m*. The effective mass *m* depends on the nature of the crystal lattice and varies with the direction of motion of the electron in the lattice, *m* may be much larger or much smaller than *m*, and it may even be negative.

Let us consider an electron moving along the *x*-axis in a crystal lattice to which an external electric field *E* has been applied. The external force acting on the electron is *eE*. Suppose that the electron gains velocity *v* over a distance *dx* in time *dt*. The resulting change in its energy (work done by the force) is given by

$$d\in = eE\,dx - eE\,v\,dt. \qquad (\therefore\ v = dx/dt)$$

We know that the velocity *v* of a particle is the same as the 'group velocity' $vg(= d\omega/dk)$ of the de Broglie waves associated with the particle ($v = vg$). Thus

$$d\in = eE\,vg\,dt. \qquad \text{... }(i)$$

Now, the Einstein-de Broglie relation is

$$\in = hv = \frac{h}{2\pi}\omega \qquad \left[\therefore v = \frac{\omega}{2\pi}\right]$$

where ω is the angular frequency associated with c. Differentiating it, we have

$$d\in = \frac{h}{2\pi}\,d\omega = \frac{h}{2\pi}\frac{d\omega}{dk}\,\mathrm{dk}. \qquad \text{... }(ii)$$

But $\qquad \dfrac{d\omega}{dk} = v_g$ (definition of group velocity).

$$\therefore \qquad d\in = \frac{h}{2\pi} v_g \, dk.$$

Comparing it with eq. (*i*), we get

$$eE\, dt = \frac{h}{2\pi} dk$$

or

$$\frac{dk}{dt} = \frac{2\pi}{h} eE. \qquad \text{... } (iii)$$

Let us write the group velocity v_g in terms of energy $\in$.

Differentiating with respect to time, we get

$$\frac{dv_g}{dt} = \frac{2\pi}{h} \frac{d^2 \in}{dt\, dk} = \frac{2\pi}{h} \frac{d^2 \in}{dk^2} \frac{dk}{dt}$$

or

$$\frac{dv_g}{dt} = \frac{4\pi^2}{h^2} \frac{d^2 \in}{dk^2} eE. \qquad \text{[by } e\ q.\ (iii)\text{]}.$$

Employing v_g = ,v again, this can be written

$$\frac{dv}{dt} = \left(\frac{4\pi^2}{h^2} \frac{d^2 \in}{dk^2} \right) eE.$$

This equation connects the force eE on the electron with the acceleration dv/dt through the proportionality factor $\frac{4\pi^2}{h^2} \frac{d^2 \in}{dk^2}$. It is same as Newton's second law (acceleration = force/mass), if we set

$$\boxed{\frac{1}{m} = \frac{4\pi^2}{h^2} \frac{d^2 \in}{dk^2}}$$

The quantity 1/m is the reciprocal of the effective mass of the electron in the crystal lattice. In other words, an election in a lattice responds to an external force precisely as a free electron would if its mass were m.

$\frac{1}{m}$ is determined by the properties of the lattice on which depends the form of the energy function $\in$. The first Brillouin zone $\left(\text{from } -\frac{\pi}{a} \text{ to } +\frac{\pi}{a}\right)$ and the part of the second Brillouin zone $\left(\text{beyond } -\frac{\pi}{a} \text{ and } +\frac{\pi}{a}\right)$, and the corresponding energy bands and gaps, of a one-dimensional lattice have already been mentioned.

The solid curve is for the electron in crystal lattice. The dashed parabola is for the free electron (moving in a uniform potential) having

$$\in = \frac{h^2k^2}{8\pi^2 m}$$

which amounts to

$$\frac{1}{m} = \frac{1}{m} \qquad \text{(for all values of } k)$$

Thus, near the centre of the first zone, where the two curves coincide ($m = m$), the lattice has little effect on the electron which responds to the applied electric field as if it were free. This is true for the electrons at the bottom of the first energy band. The applied field-force will increase the electron's energy and momentum.

As we move in either direction from the center of the first zone, the curvature $\frac{d^2 \in}{dk^2}$ deviates significantly from the parabola. It goes through zero and then becomes negative. $\frac{1}{m}$ does the same. Thus in the upper part of the first energy band electron in the lattice responds to the applied field very differently from that a free electron would do. Where $\frac{1}{m}$ is zero (or m is finite), the applied force eE causes no acceleration of the electron. The momentum gained by the electron due to *eE is* neutralised by the momentum lost to the lattice through reflection. Near the top of the band, where $\frac{1}{m}$ is negative, the applied force causes an acceleration in a direction opposite to that would be experienced by a free electron. Here the Bragg reflection is more perfect and so the momentum transferred to the lattice from the electron is larger than the momentum transferred from the applied force to the electron. That is, there is a net *decrease* in the forward momentum of the electron. *Thus a 'negative' effective mass of the electron in the lattice physically means that the electron responds to an applied force with a 'decrease' in momentum.*

In fact the current carriers near the top of an almost-filled band are the "holes" (vacant electron states) which behave as if they have positive effective mass.

At the bottom of the next energy band (for the second Brillouin zone) $\frac{1}{m}$ is positive but larger than $\frac{1}{m}$, so the applied force produces a relatively large acceleration of the electron in the lattice.

Significance : The significance of the effective mass is that all results of the free-electron theory of metals can be taken in the more realistic band theory merely by replacing the electron mass *m* by the average effective mass m at the Fermi surface. Thus the Fermi energy in a metal is given by

$$\in_F = \frac{h^2}{2m^*}\left(\frac{3N}{8\pi V}\right)^{2/3}$$

where *N/V* is the density of valence electrons.

For example, the electronic specific heat is inversely proportional to $\in_F$ and therefore directly proportional to *m*. Metals such as Co, Ni and Pt which have high *m/m* ratios (≈ 10) have high electronic specific heats.

PROBLEMS

1. ***In germanium the energy gap is about 0.75 eV. Show that the crystal behaves as a transparent medium only for light of wavelength above 16533A. (h = 6.6 × 10^{-34} joule-sec, c = 3.0 × 10^8 meter/sec).***

Solution : The energy of 16533-Å photons is

$$E = \frac{hc}{\lambda}$$

$$= \frac{(6.6 \times 10^{-34} \text{ joule-sec})(3.0 \times 10^8 \text{ meter/sec})}{16533 \times 10^{-10} \text{ meter}}$$

$$= 1.2 \times 10^{-19} \text{ joule}$$

$$= \frac{1.2 \times 10^{-19} \text{ joule}}{1.6 \times 10^{-19} \text{ joule/eV}} = 0.75 \text{ eV.}$$

The photon energy for light of wavelength above 16533A will be less than 0.75 eV which is the energy gap in germanium. Hence these photons will not be absorbed and the light will be transmitted through the crystal.

All photons for light below 16533 Å will have an energy more than the energy gap of 0.75 eV and so they will be absorbed in exciting the electrons from the valence band to the conduction band. Hence the crystal is opaque for light below 16533 Å.

2. ***Calculate the energy gap of a crystal which is transparent only for light of wavelength greater than 12345 Å.***

[**Ans.** 1.0 eV.]

3. ***An insulator has an optical absorption only for wavelengths shorter than 1800 Å. Find the width of the forbidden band for the insulator. What is the order of magnitude of the forbidden gap in a semiconductor ?***

Solution : The energy of 1800-Å photons is

$$E = \frac{hc}{\lambda} = 6.9 \text{ eV}. \quad \text{(Calculate as in prob.1)}$$

Light of wavelength longer than 1800 Å has photons of energy less than 6.9 eV. These photons are not absorbed by the electrons, which are therefore unable to cross the forbidden band. Hence the energy width of the forbidden band is 6.9 eV.

The forbidden gap of a semiconductor is of the order of leV.

4. The energy gap in silicon is 1.1 eV and in diamond it is 6 eV. State the transparency of these substances to visible light.

Solution : Let us calculate the wavelength of light corresponding to photon energies of 1.1 and 6 eV.

$$\lambda(1.1 \text{ eV}) = \frac{hc}{E}$$

$$= \frac{(6.6 \times 10^{-34} \text{ joule} - \text{sec})(3.0 \times 10^{8} \text{ meter/sec})}{1.1 \times 1.6 \times 10^{-19} \text{ joule}}$$

$$= 11 \times 10^{-7} \text{ meter}$$

$$= 11000 \text{ Å}.$$

Similarly,

$$\lambda \text{ (6 eV)} = 2100\text{Å}.$$

Thus, silicon is transparent only to radiation of $\lambda \geq 11000$ Å; since it absorbs photons of shorter wavelength, and so it is opaque to visible light. Diamond is transparent to radiation of $\lambda \geq 2100$ A, so it is transparent to visible light.

5. ***The energy gaps for Si, Ge and Ag are 1.1, 0.7 and 0 eV respectively. Find the wavelength of electromagnetic radiation to which these solids are opaque. (h = 6.6 × 10^{-34} joule-sec. (c = 3.0 × 10^{8} meter/sec).***

[**Ans.** Si : $\lambda = 0 - 11000$ Å, Ge : $\lambda = 0.18000$ Å, Ag : $\lambda = 0 - \infty$.]

6. ***A metallic conductor carries a current of 10 amp. Its area of cross-section is 1.0 cm^2. The specific resistance of the conductor is 1.7 × 10^{-8} ohm-cm. Calculate the electric field inside the conductor.***

Solution : The current density j is given by

$$j = \frac{i}{A} = \frac{1.0 \text{ amp}}{1.0 \text{ cm}^2} = 1.0 \text{ amp/cm}^2.$$

The resistivity (specific resistance) ρ of a metal is defined as

$$\rho = \frac{E}{j}.$$

Therefore, the electric field inside the metal is

$$\begin{aligned} E &= \rho j \\ &= (1.7 \times 10^{-6} \text{ ohm-cm})\,(1.0 \text{ amp/cm}^2) \\ &= 1.7 \times 10^{-6} \text{ volt/cm} \\ &= 1.7 \times 10^{-4} \text{ volt/meter.} \end{aligned}$$

7. ***Find the average drift velocity of electrons in a copper conductor with a cross-sectional area of 10^{-6} m^2 carrying a current, of 4 A. The atomic weight of copper is 63.6 and the density is 8.9 gm/cm^3. Take Avogadro's number N_A = 6.02 × 10^{23}.***

Solution : The number of copper atoms per meter^3 of the conductor is

$$\frac{N_A \rho}{M} = \frac{6.02 \times 10^{23} \text{ atoms/gm-atom} \times 8.9 \text{ gm/cm}^3}{63.6 \text{ gm/gm-atom}}$$

$$= 8.42 \times 10^{23} \text{ atoms/cm}^3$$

$$= 8.42 \times 10^{28} \text{ atoms/m}^3.$$

Assuming that there is one free electron per atom, the electron density is

$$n = 8.42 \times 10^{28} \text{ electrons/m}^3.$$

If j be the current density and vd the drift velocity, then we have

$$j = ne\, v_d$$

$$\therefore \qquad vd = \frac{j}{ne} = \frac{i/A}{ne}$$

$$= \frac{4 \text{ amp}/10^{-6}\text{ m}^2}{\left(8.42 \times 10^{28} \text{ electrons}/\text{m}^3\right)\left(1.6 \times 10^{-19} \text{ coul}/\text{electron}\right)}$$

$= 3.0 \times 10^{-4}$ m/sec.

8. *What would be the mobility of electrons in copper if there are 9×10^{28} valence electrons per m^3 and the conductivity of copper is 6×10^7 mho/m ?*

Solution : The metallic conductivity is given by

$$\sigma = ne\mu ,$$

where n is the free-electron density in the metal and μ is the mobility of electrons. Thus

$$\mu = \frac{\sigma}{ne}$$

Here $\sigma = 6 \times 10^7$ mho/m $= 6 \times 10^7$/ohm-m.

$$\therefore \qquad \mu = \frac{6 \times 10^7 \text{ /ohm-m}}{\left(9 \times 10^{23}/\text{m}^3\right)\left(1.6 \times 10^{-19} \text{ coul}\right)}$$

$$= 4.16 \times 10^{-3} \text{ m}^2\text{/ohm-coul}$$

$$= 4.16 \times 10^{-3} \text{ m}^2\text{/volt-sec.}$$

9. *The Fermi energy of copper is 7 eV. Calculate the average distance (mean free path) travelled by the conduction electrons between collisions. Take the conductivity of copper as 6×10^7/ohm-m and the concentration of the valence electrons as $8.5 \times 10^{28}/m^3$. ($m = 91 \times 10^{-31}$ kg, $e = 1.6 \times 10^{-19}$ coul and 1 eV $= 1.6 \times 10^{-19}$ joule).*

Solution : The electrons which effectively contribute to the electrical conductivity are those with energies near the Fermi energy $\in F$. The velocity of these electrons is given by

$$v_F = \sqrt{\left(\frac{e \in_F}{m}\right)}$$

$$= \sqrt{\frac{2 \times \left(7 \times 1.6 \times 10^{-31}\ joule\right)}{9.1 \times 10^{-31}\ kg}}$$

$$= 1.6 \times 10^6 \text{ m/s.}$$

Now, the metallic conductivity σ is, in terms of mean free path λ, given by

$$\sigma = \frac{ne^2\lambda}{2mv_F}$$

so that $$\lambda = \frac{2mv_F\ \sigma}{ne^2}$$

Substituting the given values, we get

$$\lambda = \frac{2 \times \left(9.1 \times 10^{-31}\ \text{kg}\right) \times \left(1.6 \times 10^5\ \text{m/s}\right) \times \left(6 \times 10^7\ /\text{ohm} - \text{m}\right)}{8.5 \times 10^{28}\ \text{m}^3 \times \left(1.6 \times 10^{-19}\ \text{coul}\right)^2}$$

$$= 8 \times 10^{-8}\ m = 800\text{Å}.$$

Since the ions that make up the lattice of solid copper are 2.6 Å apart, a free electron travels past about 300 of them, on the average, before a collision occurs.

16

Photoconductive Device

Photoconductor

When radiation falls upon a semiconductor, its conductivity increases. This is 'photo conductive' effect and is the basis of a photoconductor.

A photoconductor is a device that detects optical signals. In its simplest form, it consists of a large-surface slab of intrinsic or almost intrinsic semiconductor, in bulk or thin-film form,

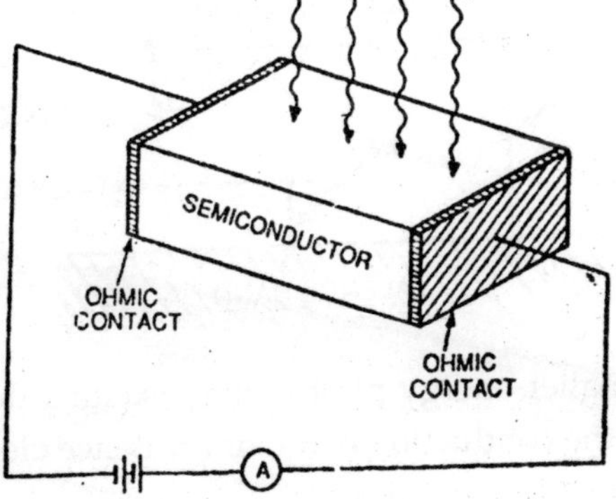

with ohmic contacts affixed to opposite ends. A commercial photo conductor, known as photo conductive cell, has a layer of cadmium sulphide (Cd S) containing a small amount of SB or In impurity on its sensitive surface. When light falls the surface of the photo conductor, the current in the circuit increases depending on the intensity of the incident light.

Working Principle : The conductivity of a material is proportional to the concentration of charge-carriers in it. In a semiconductor there are thermally-generated electron-hole pairs. When radiation falls on it, bivalent bonds are broken and additional pairs are created. These increased charge-carriers increase the conductivity, or decrease the resistance of the semiconductor. Hence the photo conductor is also known as 'photoresistor'.

The photo-generation of electron-hole pairs is shown in Fig. below by means of the energy band diagram of a semiconductor having both donor and acceptor impurities. When photons of sufficient energies fall on the semiconductor, there are two possibilities :

(i) A photon of energy greater than the energy gap ϵ_g may excite a valence-band electron into the conduction band, thus creating an electron in the conduction band and a hole in the valence band. This is called 'intrinsic excitation,

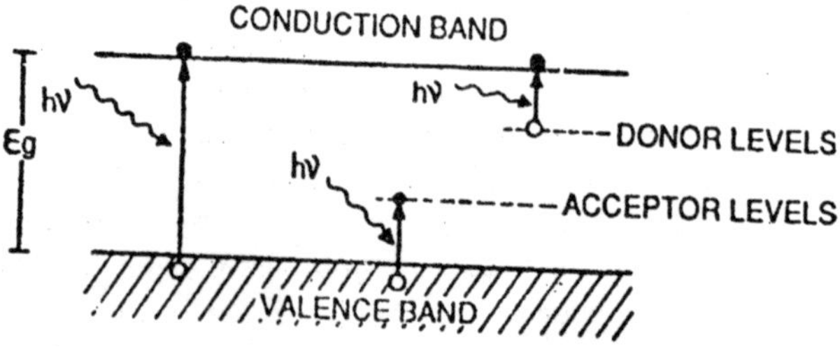

(ii) A smaller-energy photon may excite a donor electron into the conduction band, or a valence electron may go into acceptor level. These transitions which also generate electron-hole pairs are known as 'impurity or extrinsic excitations'. Since the density of states in the conduction and valence bands is larger than the density of impurity states, the photoconductivity is due mainly to intrinsic excitations.

The minimum energy of a photon required for intrinsic excitation is the forbidden-gap energy ϵ_g of the semiconductor. The long-wavelength threshold of the semiconductor corresponding to ϵ_g is given by

$$\lambda_{max} = \frac{hc}{\epsilon_g} = \frac{1.24}{\epsilon_g\,(eV)}\ \mu\ \text{(micron)}$$

(For Si, $\epsilon g = 1.1$ eV and $\lambda_{max} = 1.13\ \mu$; for Ge, $\epsilon_g = 0.7$ and $\lambda_{max} = 1.77\ \mu$ at room temperature). For wavelengths shorter than λ_{max} , the incident radiation is absorbed by the semiconductor, and electron-hole pairs are created. (In fact, the long-wavelength limit is slightly greater than the value of λ_{max} calculated above, because of the impurity excitation). At longer A, the semiconductor becomes transparent.

The carriers generated by photoexcitation move under the applied p.d. across the photoconductor, reach the ohmic contacts at the ends of the semiconductor and constitute the photo conductive current which is added to the small circuit current *(i.e.* current when no radiation is falling).

Uses : Photo-conductive devices, or cells, are used in industry, photography and light-intensity measurements. The most commonly used cell is the cadmium sulphide (Cd S) photo-conductive cell which is excellently sensitive in the visible range and gives a maximum response at about 5800 A. It can handle power levels of several watts and so directly operate a relay (without requiring an intermediate amplifier) or control the opening of a camera lens.

Other photo-conductive cells are available for specific purposes. A lead sulphide (Pb S) or indium-antimony (In Sb) cell is used for infrared-detection or infrared absorption measurements. A selenium cell is particularly sensitive in the blue region.

Photo conductors are relatively simple to construct, can be made in poly-crystalline material and so are also cheap. They are, however, relatively slow and require an external voltage source.

p-n Photodiode : It is a junction-type photoconductor having several advantages over an ordinary bulk-type photoconductive cell. It consists of a reverse-biased *p-n* junction embedded in plastic. The upper surface across the junction is open to radiation, while the remaining sides of the plastic are either painted black or enclosed in a metallic case. The entire unit is extremely small, of the order of a 0.1 inch size. The mechanism of current flow due to radiation can be explained by means of energy band diagram of the *p-n* junction. In the absence of radiation, an almost constant current (independent of reverse bias which is of the order of a few tenths of a volt) is obtained This is the reverse saturation current due to the thermally-generated minority carriers (electrons in the *p*-type region and holes in the *n*-type region).

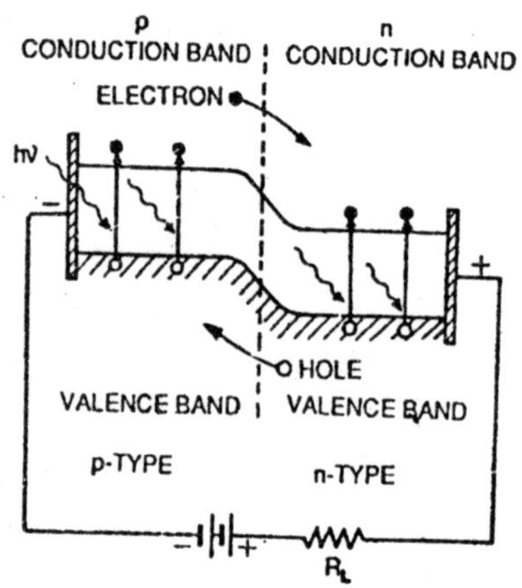

When radiation falls upon the surface, additional electron-hole pairs are formed. The photo- excited electrons in the conduction band of the *p*-type move across the junction (along with the thermally-generated minority electrons) to the *n*-side and add to the current. Similarly, the holes produced in the valence band of the *n*-type (due to electrons being photo-excited into the conduction band) move across the junction to the *p*-side and add to the current. The photo-conductive current so produced varies almost linearly with the radiation flux.

The *p-n* photo diode can operate at frequencies of the order of 1 MHz, in contrast to the slow bulk-type photo conductor which is limited to a frequency range from 100 to 5000 Hz. Hence the photodiodes are extensively used in high-speed reading of computer punched cards, light-detection systems, light-operated switches, counting of objects interrupting a light-beam, etc.

Solar Cell

A solar cell is a semiconductor device (*p-n* junction) which converts solar energy directly into electrical energy. It is based on the phenomenon of photovoltaic effect.

Construction : A schematic representation of a silicon solar cell is shown in Fig. below. It consists of a shallow *p-n* junction formed at a short distance below the surface of a silicon crystal. There is an ohmic contact at the top in the form of a comb-like structure which covers above 7% of Si area, and an ohmic contact at the bottom which covers the entire bottom area. The top carries an anti-reflection coating. When sunlight falls on the top surface, it is absorbed mostly near the surface where the junction is located. This results in an electrical p.d. generated between the top and the bottom of the cell which can deliver power to an external circuit. The metallic comb-like structure prevents undue increase of internal resistance of the cell because of the thinness of the *n*-type Si layer.

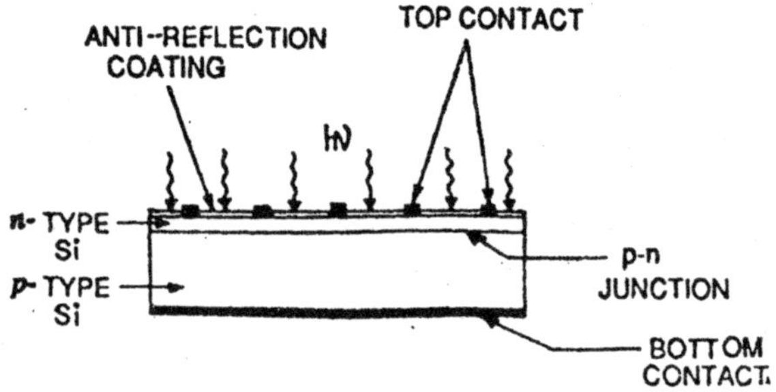

Principle : When a beam of photons is incident on the depletion region of an unbiased *p-n* junction, each absorbed photon creates an electron and a hole. As shown in the energy-band diagram of the junction in Fig. (a) below, the absorption of photon causes an electron in the valence band to move to the conduction band, and a hole is left in the valence band. The built-in (internal) electric field E_4 in the depletion region forces the hole towards the *p*-type region and the electron towards the *n*-type region. The two sides of the junction thus become oppositely charged so that a forward voltage is generated across the junction. This voltage drives a current around an external circuit, and the junction acts as a cell. The equivalent circuit of the solar cell is shown in Fig. (b) below, in which R_L is the external load.

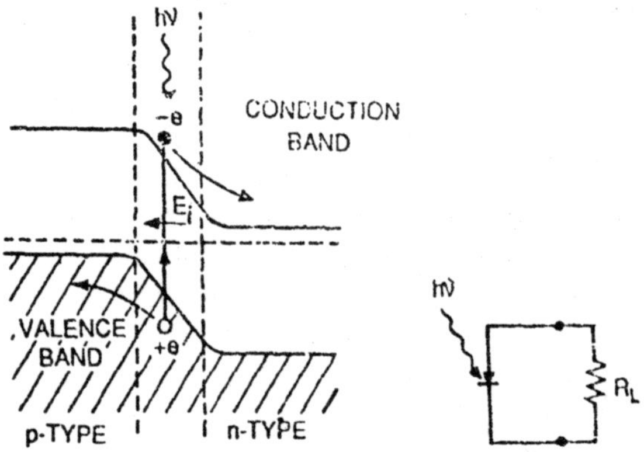

(a) (b)

Uses : The commercial cells are made from single Si crystals having efficiencies around 15%, open-circuit voltages of nearly 0.6 volt and areas of up to 40 cm^2. They are arranged in large-area arrays and connected in series and parallel to obtain useful load voltages and currents. This makes solar cells a comparatively expensive energy source. Hence their use is limited to special purposes; such as in space vehicles where the small mass of the solar batteries is very important, or inaccessible places on earth where automatic equipment must keep operating unattended.

Efforts are being made to reduce the cost of solar cell by shifting from single Si crystal to cheaper polycrystalline Si or amorphous Si, or by using cheap plastic lenses to concentrate sunlight on a small-area (and hence cheaper) high-quality cell.

Semiconductor Injection Laser

A laser, meaning Light Amplification by Stimulated Emission of Radiation, is a device that converts electrical energy into optical radiation in the form of an intense, monochromatic, directional and highly coherent beam. Like ruby crystal and He-Ne gases, semiconductor junction diodes can also be used as laser. The first semiconductor laser was made from gallium arsenide, Ga As.

Principle : A semiconductor has a small energy gap ϵ_g (≈ 1 eV) between its valence band having energy states occupied by electrons and the empty conduction band. At room temperature, however, some of the valence electrons acquire thermal energy greater than ϵ_g and cross over into the conduction band, leaving behind holes in the valence band. If a light photon of energy greater than ϵ_g happens to interact with the electrons, one of the two processes may occur : (*i*) the photon may be absorbed by a valence-band electron which would be excited to the conduction band leaving behind a hole in the valence band (Fig. *a*) ; (*ii*) the photon may stimulate an already excited conduction-band electron which would drop to the valence band, emitting a fresh photon in coherence with the stimulating photon (Fig. b). Which of the two processes is more probable depends upon whether most of the electrons are in the valence band or in the conduction band.

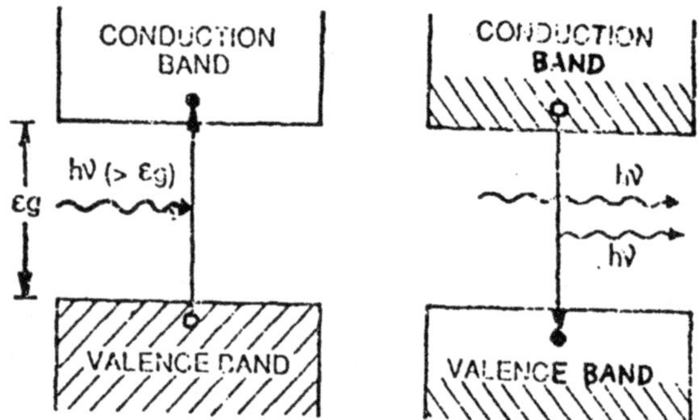

Under ordinary conditions of thermal equilibrium the number of electrons in higher-energy states (conduction band) is much smaller than in lower-energy states (valence band), so that there is little stimulated emission compared with absorption. *If, however, by some means a large number of electrons are made available in high-energy states, stimulated emission is promoted.* This situation is called "population inversion". Once population inversion is achieved, the photons go on multiplying by repeated stimulated emissions and a strong coherent beam of light emerges from the semiconductor.

Achievement of Population Inversion : Population inversion can be achieved near a *p-n* junction having high doping densities and forward currents. The large number of injected carriers creates a region near the junction where there is a very large number of electrons in the conduction band together with a very large number of holes in the valence band, *i.e.*, a population inversion.

Basic Structure : The basic structure of a gallium arsenide *p-n* junction, used as an injection laser, is shown in Fig. below. The pair of parallel planes perpendicular to the plane of the junction is polished, while the two remaining sides of the diode are roughened. When a *forward* bias is applied to the laser diode, a current flows. The injected electrons move from the *n*-side to the *p-side* and the holes from the *p*-side to the *n*-side. As the electrons and holes re-combine, photons are emitted. These photons are re-absorbed or radiated away. This is the spontaneous emission which occurs in all directions at low current. As the current is increased, eventually a threshold current is reached at which the emitted photons stimulate the emission of more photons. These photons are internally reflected several times at the polished walls, stimulating more and more photons, all coherent with them. When the photon beam becomes sufficiently intense, it emerges out from the junction.

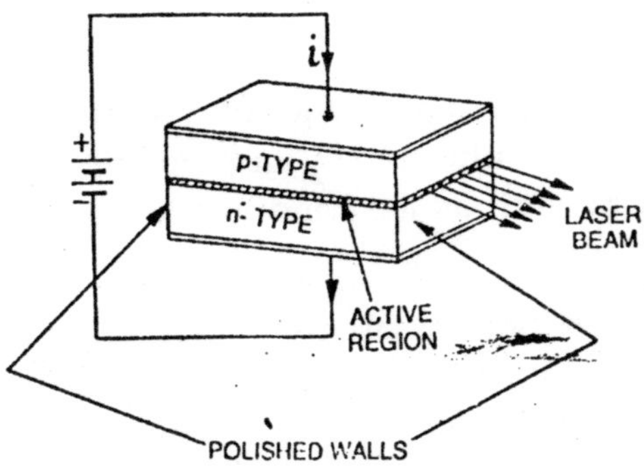

The main difficulty with this Ga As laser is the high threshold current density ($\approx 10^5$ A/cm^2) at room temperature. Hence this laser could be operated only at low temperatures at which the required current density is lower. Heterostructure lasers have now been built which operate at room temperature at moderate current densities.

Comparison with Other Lasers : Semiconductor lasers are similar to other lasers in that the emitted radiation is intense, monochromatic, directional and highly coherent, both spatially and temporally. However, in some respects, semiconductor lasers differ from other lasers :

(i) In semiconductor lasers, the electron transitions are associated with the band structure of the materials, whereas in other lasers the transitions take place between discrete energy levels. Therefore, the radiation from a semiconductor laser is less monochromatic (has a wavelength range of few angstroms) than that from other lasers (which have a range of only a fraction of an angstrom).

(ii) Because the active region in a semiconductor laser is very narrow ($\approx 1\ \mu m$ thickness), the divergence of the laser beam is considerably larger than in case of other lasers.

(iii) A semiconductor laser is very compact in size, less than 1 mm.

(iv) The spectral characteristics of a semiconductor laser are strongly influenced by the properties of the junction medium, such as the band gap and refractive index variations.

(v) In the *p-n* junction laser, the laser action is produced by passing a forward current through the junction. Hence the emitted laser light can be modulated into light pulses simply by modulating the current.

(vi) Semiconductor lasers have very short photon life times. Therefore, modulation at high frequencies can be achieved.

Applications : Lasers have found a wide use in many fields, such as precision measurements of long distances and three-dimensional lensless photography (holography). The distance to the moon, for example, has been determined to an accuracy of 15 cm.

In medicine, laser beams are used as finest surgery tools which seal the blood vessels as they cut. They are also used in the treatment of tumors and retinal detachment.

In industry, laser beams are used as machine tools for drilling extremely fine holes in diamonds, teeth, paper clips and even in human hairs. They are used also for cutting materials, from fabrics to ceramics and metals. Welding on a microscopic scale, such as in integrated circuits, is taken up by means of laser beams.

The semiconductor laser, because of its compact size and capability for high-frequency modulation, is the best light source for optical-fiber communication.

17

Microelectronic Circuits

Integrated Circuits

We are familiar with the 'discrete circuits' consisting of separately manufactured active and passive circuit components, externally inter-connected by wires. The circuits are called 'discrete', because each component of the circuit is discrete from the others. These circuits occupy large space, and have a number of joints which make them somewhat unreliable.

To meet the military requirement of miniature (mini) electronic equipment, a new branch of electronics called 'microelectronics' originated in the late 1950s. It deals with microelectronic circuits called as "integrated circuits", abbreviated as ICs. In an IC all components like resistors, capacitors, diodes, transistors, etc. are fabricated on a monolithic (single) semiconductor chip. Thus an integrated circuit is a packaged electronic circuit which has the advantages of high reliability, small physical size, low cost, and low power consumption.

The individual components of an IC can, however, neither be removed nor be replaced because each one of them is an integral part of the same semiconductor chip. A typical size of a semiconductor chip is 50 mils × 50 mils × 5 mils (1 mil = 0.001 inch).

Types of Integrated Circuits

Monolithic ICs : A monolithic IC is one in which all circuit components and their interconnections are formed on a single semiconductor chip. Monolithic circuit components include resistors, capacitors, transistors (BJTs and FETs).

Film ICs : A film IC is one in which the circuit components are formed on a substrate. The components include resistors, capacitors and thin-film transistors.

Hybrid ICs : A hybrid IC is a combination of two or more ICs or one IC and several discrete components.

Multi-chip ICs : In a multi-chip IC, the circuit components are fabricated on separate chips which are attached to a substrate and interconnected as if they were discrete components. These circuits have limited use.

Integrated Circuits (ICs) : A (monolithic) integrated circuit consists of a single silicon chip in which both active and passive components have been diffused, and interconnected by aluminium metallization.

Fabrication of a Monolithic IC : With the exception of inductance, all the electronic circuit components (resistors, capacitors, diodes, junction and field-effect transistors) can be formed in semiconductor form. Therefore, the fabrication of all ICs involves the same chain of processes, which are as follows:

1. Wafer Preparation,
2. Epitaxial Growth.
3. Isolation Diffusion.

4. Base Diffusion.
5. Emitter Diffusion.
6. Pre-ohmic Etching and Metallization.
7. Checking and Dicing.
8. Mounting and Packaging.

Wafer Preparation : First of all, a single and pure *p*-type silicon crystal is grown and cut into wafers (thin slices) which are cleaned and polished to a mirror finish. The typical thickness of a finished wafer is 5 mils and its resistivity is 10 Ω-cm. This lightly-doped *p*-type wafer provides the base or 'substrate' on which the circuit components (resistor, transistor; etc.) are to be built.

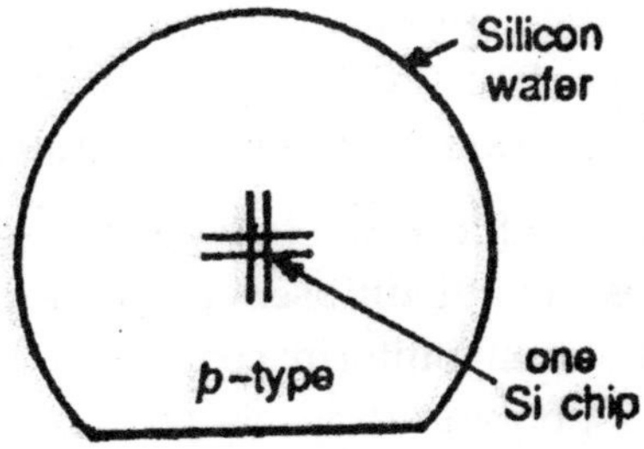

A silicon wafer, about 1.5 inches in diameter, contains several hundreds of chips. In the processes which will follow, exactly identical circuits are produced simultaneously on all the chips. After the final process, the individual chips are separated by cutting.

Epitaxial Growth : The active and passive components are built within a thin *n*-typs 'epitaxial layer' on top. Therefore, an *n*-type layer of silicon, typically 1 mil thick, is grown on the *p* type substrate by placing the wafer in a furnace at 1200°C and introducing as gas containing phosphorus (donor impurity), as shown in Fig. below. The resistivity of this *n*-type layer ranges from. 0 1 to 0 5 .Ω-cm.

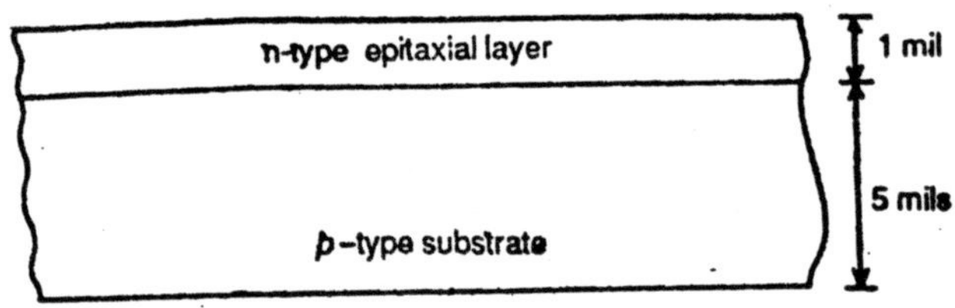

Types of Diffusion

Isolation Diffusion : The n-type epitaxial layer is isolated into islands, so that each component may be formed on a separate island. This is done by diffusing p-type impurity (boron) through the n-type epitaxial layer to the p-type substrate by a series of steps common in all diffusion processes :

A thin layer (~ 1 micron thick) of silicon dioxide (SiO_2) is formed on the n-type epitaxial layer by exposing it to oxygen and heating to about 1000°C (Fig. a). SiO_2 has the property of preventing the diffusion of impurities through it.

Now, in order to form selective openings in SiO_2 through which impurities may be diffused, a photoetching method is used. For this, a thin uniform coating of a photosensitive emulsion, called 'photoresist' is laid on the SiO_2 layer (Fig. b).

Then a large black-and-white layout of the desired pattern of openings is made and reduced photographically. This negative is placed as a "mask" over the photoresist which is then exposed to ultraviolet light (Fig. c).

The photoresist under the transparent regions of the mask becomes polymerized. The mask is now removed, and the wafer is "developed" by a chemical (like trichloroethylene) which dissolves the unexposed (unpolymerized) portions of the photoresist coating and leaves the surface pattern as in Fig. (d).

The wafer is immersed in an etching solution (hydrofluoric acid) which removes SiO_2 from the areas not protected by the photoresist (Fig. e).

The photoresist is removed completely by scrubbing with heated solvents (Fig. f). The wafer is now ready for the isolation diffusion. The remaining SiO_2 serves as a mask for the diffusion of *p*-type impurity (boron).

The wafer is placed in a "boat" and passed through a' furnace containing boron gas. The *p*-type impurity diffuses into the wafer through the openings in SiO_2, turning the *n*-type material into a p^+–type channel down to a depth extending to the *p*-type substrate.

Thus the *n*-type epitaxial layer is isolated into islands called 'isolation islands' or 'isolated regions' resting on the *p*-type substrate under the SiO_2 layer (Fig. g). Their purpose is to provide electrical isolation between different circuit components The *p*-type substrate is always held at a negative potential with respect to the isolation islands so that the *p-n* junctions are reverse-biased, otherwise the isolation will be lost.

The concentration of the *p*-type impurity atoms in the regions between isolated islands *i.e.* in p+-type channels, is much higher (and hence indicated as p^+) than that in the *p*-type substrate to prevent any connection between two isolated islands.

The individual circuit components are now built within these isolated islands. Let us consider the fabrication of an *n-p-n* transistor and a resistor in two adjoining islands.

Base Diffusion : A part of the *n*-type island itself provides the collector for the *n-p-n* transistor. The *p*-type base of the transistor is diffused into the collector. At the same time, diffusion of the resistor takes place in the adjoining island. For this a complete new layer of SiO_2 is formed over the wafer and all the above processes are repeated using a different mask so as to create a pattern of openings shown in

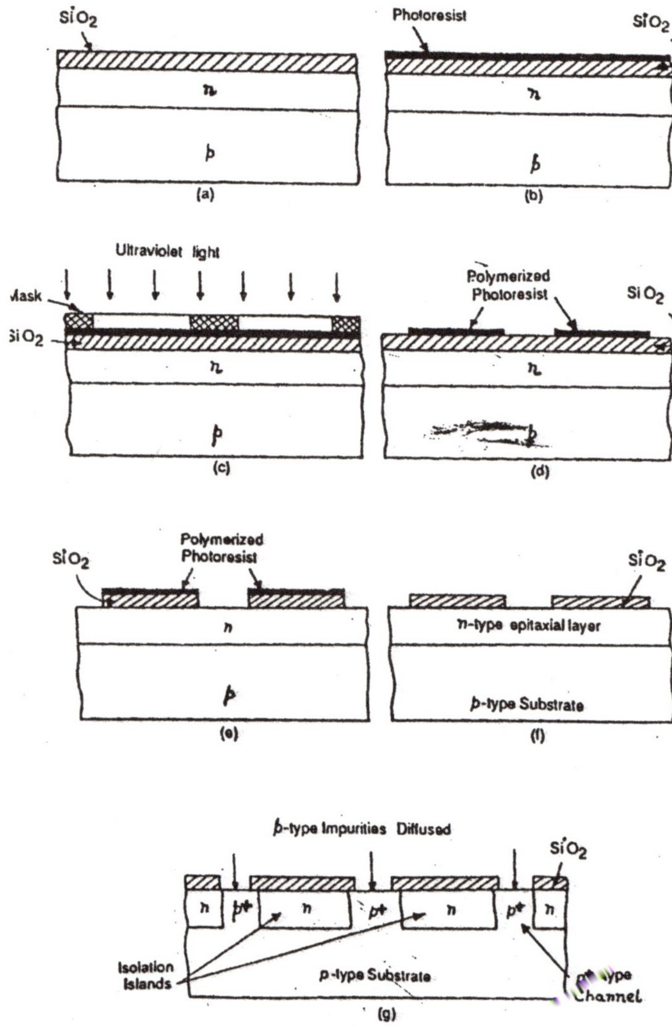

Fig. below. The p-type impurity (boron) is diffused through these openings. In this way are formed the transistor base region and the resistor. The depth of this diffusion is kept controlled (by controlling the time of diffusion) so that is does not penetrate to the substrate. The resistivity of the base layer is generally much higher than that of the isolation regions.

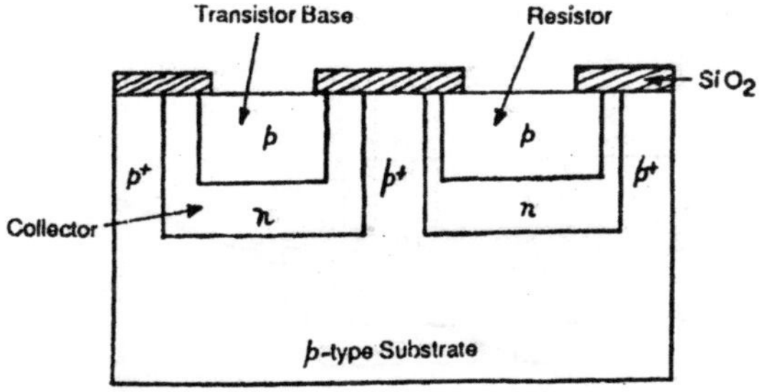

Emitter Diffusion : A layer of SiO_2 is again formed over the entire surface and the masking and etching processes are repeated to create an opening in the *p*-type base region, as shown in Fig. below. Now, *n*-type impurity (phosphorus) of heavy concentration (called n^+) is diffused through this opening for the formation of the transistor emitter. No diffusion takes place in the adjoining island, since the resistor is complete.

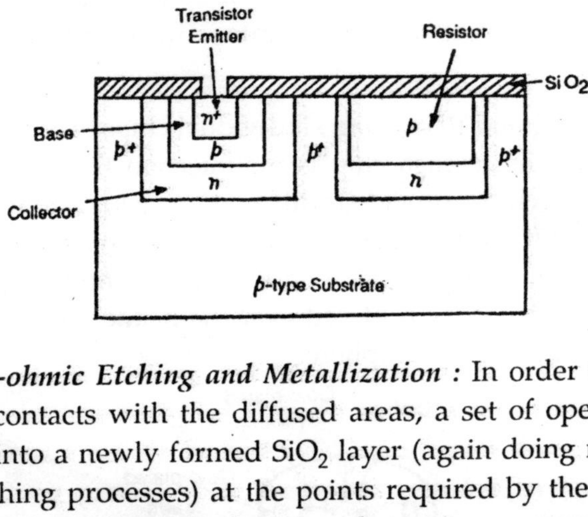

Pre-ohmic Etching and Metallization : In order to make ohmic contacts with the diffused areas, a set of openings is made into a newly formed SiO_2 layer (again doing masking and etching processes) at the points required by the desired circuit. Inter connections between the various components of the integrated circuit are then made by aluminium metallization. For this a thin coating of aluminium is deposited over the entire wafer by a vacuum evaporation of aluminium, and the undesired aluminium areas are then etched away. This leaves the desired pattern of ohmic contacts and interconnection, as shown Fig. below in which an *n-p-n* transistor and a resistor have been fabricated on chip. The circuit symbol is also shown.

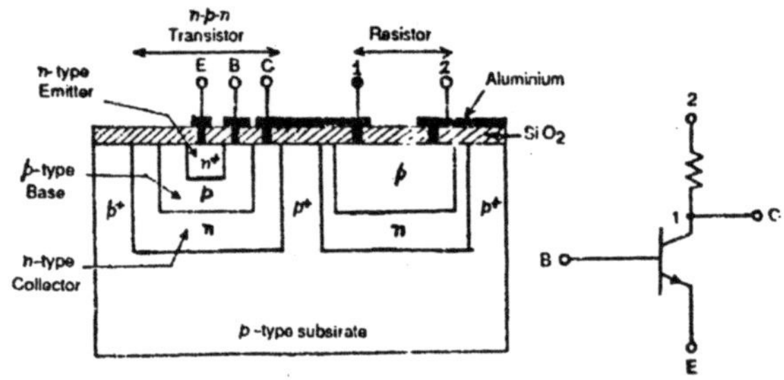

Checking and Dicing : During processing, a large number (several hundred) of ICs are manufactured on a single wafer. After the metallization process has been completed, each IC on the wafer is checked electrically for proper functioning, and the faulty circuits are marked. The wafer is then scribed with a diamond point and separated into individual chips (or dice) containing the integrating circuits. The faulty chips are discarded.

Mounting and Packaging : Individual chips are very small and brittle. Hence each chip is mounted on the gold-plated leads and the chip is provided with 'bonding pads'. Connections between the IC and the package leads are done by aluminium wires from the bonding pads on the chip to the leads on the header. Finally, a cap is placed over the header and the IC is seeded in an inert atmosphere.

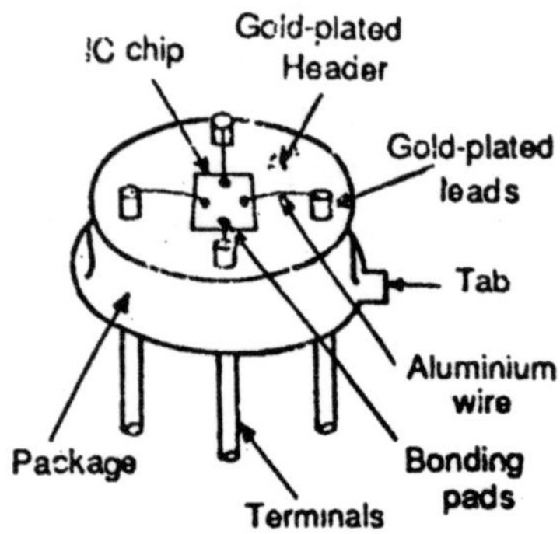

Applications of ICs : Integrated circuits have the advantages of small size, light weight, low cost (because thousands of complex units are fabricated simultaneously), low power consumption, and high reliability. Therefore, they are frequently used in space vehicles, hearing aids and all types of computers.

ICs are of Two Types : linear and non-linear. The linear ICs are used in power amplifiers, high-frequency amplifiers, differential operational amplifiers, voltage regulators and in analog computer circuits.

The non-linear ICs are used in great quantities in digital computers to perform switching functions in logic gates and memory units. The small size of MOS components has led to "large scale integration" (LSI) in which thousands of components are created on a single chip. Such ICs are used, for example, in pocket calculators.

Bibliography

Anotola, S.A. : *Advanced Physics,* Longman, London, 1968.

Atherton, J.C. : *Thermal Physics,* Butterworths, London, 1987.

Atiyah, M. : *The Geometry and Physics of Knots,* Cambridge, New York, 1991.

Bemier, G. and Kinet J.M. : *Thermal and Statistical Physics,* C.R.C. Press, Boca Raton, 1981.

Benjamin, W.A. : *Problems in Physics,* Academic Press, New York, 1970.

Bowling, D.J.F. : *Basic Physics,* Chapmann and Hall, London, 1976.

Caro, D. E. and Others : *Modern Physics : An Introduction to Atomic and Nuclear Physics,* Edward Arnold, London, 1985.

Cartwright, H.R. : *Elements of Thermal and Statistical Physics,* Oxford University Press, London, 1966.

Clinton, Case M. : *Physical Principles of Flow in Unsaturated Porous Media,* Oxford, New York, 1994.

Crafts, A.S. : *Phloem Transport in Plants,* W.H. Freeman, San Francisco, 1973.

Douglas, J.S. : *Advanced Guide to Modern Physics,* Pelham Books, London, 1976.

Epstein, E. : *Statistical Physics,* John Wiley and Sons, New York, 1972.

Esau, K. : *Physical Geology,* John Wiley and Sons, New York, 1965.

Evans, L.T. : *Thermal Analysis of Materials,* W.A. Benjamin, California, 1975.

Feynman and Others : *The Feymman Lectures on Physics,* Add. Wesley, London, 1989.

Feynman, R. P. and Others : *Elementary Particles and the Laws of Physics,* Cambridge, New York, 2002.

Ftzler M.E. : *Plant Lectins : Molecular and Biological Aspects,* Academic Press, New York, 1985.

Fujimoto, Minoru : *The Physics of Structural Phase Transitions,* Springer, London, 2000.

Galliard, T. and Mercer, E.I. : *Recent Advances in the Chemistry and Biochemistry of Plant Lipids,* Academic Press, New York, 1975.

Green, J.R. : *Modern Physics,* MacMillan, London, 1911.

Haasen, Peter : *Physical Metallurgy,* Cambridge, New York, 1996.

Halevy, A.H. : *Handbook of Physics,* C.R.C. Press, Boca Raton, 1985.

Hopkins, C.R. : *Structure and Function of Cells,* W.B. Saunders Co., London, 1978.

Hornberger, George M. : *Elements of Physical Hydrology,* Johns Hopkins, New York, 1998.

Jorgensen, T. P. : *The Physics of Golf,* Springer, New York, 1999.

Kirkby, E.A. : *Mathematical Physics,* The University of Leeds, Leeds, 1970.

Kuhn, Karl F. : *Basic Physics,* John Wiley, London, 1996.

Kumar, Arun : *Problems in Physics,* Atlantic Publishers, New Delhi, 2003.

Landshoff, P. and Others : *Essential Quantum Physics,* Cambridge, New York, 2001.

Leo, William R. : *Techniques for Nuclear and Particle Physics Experiments : A How to Approach*, Springer, London, 1994.

Lipson, S. G. and Others : *Optical Physics*, Cambridge, New York, 1995.

Lockett, Keith : *Physics in the Real World*, Cambridge, New York, 1995.

Lui, Lam : *Introduction to Nonlinear Physics*, Springer, New York, 1997.

Main, Lain G. : *Vibrations and Waves in Physics*, Cambridge, New York, 1998.

McKee, H.S. : *University Physics*, Oxford University Press, London, 1962.

Meidner, H. and Sheriff, D.W. : *Water and Plants*, Blackie and Sons, Glasgow, 1976.

Mengel, K. and Kirkby, E.A. : *Thermal and Statistical Physics*, Praeger Publishers, New York, 1974.

Nabapro, F. R. N. and Others : *The Physics of Creep*, T & F, New York, 1995.

Negele, J. W. and Others : *Advances in Nuclear Physics*, Plenum, London, 1994.

Palladin, V.I. : *Physical Metallurgy*, Arihant Publishers, Jaipur, 1988.

Plischke, M. and Others : *Equilibrium Statistical Physics*, World Scientific, New York, 1994.

Rabinowitch, E. : *Thermodynamics and Statistical Physics*, John Wiley, New York, 1969.

Rawn, J.D. : *Encyclopaedia of Modern Physics*, Harper and Row, New York, 1983.

Roberts, J.A. and Tucker, G.A. : *Physics and Chance*, Oxford, New York, 1999.

Rossi, Paolo : *The Birth of Modern Science*, Blackwell, New Jersey, 2001.

Ruhland, W. : *Encyclopaedia of Thermodynamics Physics*, Springer-Verlag, Berlin, 1961.

Sabnis, D.D. and Hart, J.W. : *Encyclopaedia of Stastical Physics*, Springer Verlag, Berlin, 1982.

Sagdeev, R. Z. : *Nonlinear Space Plasma Physics*, AIP, New Jersey, 1993.

Salisbury, F.B. and Ross, C.W. : *Thermal Physics*, Woodworth Publishing Co., California, 1967.

Schein, L. B. : *Electro-Photography and Development Physics*, Springer, London, 1992.

Slatyer, R.O. : *The Physics of Structural Phase Transitions*, Academic Press, New York, 1967.

Srivastava, H.S. : *Elements of Physics*, Rastogi Publications, Meerut, 1990.

Thimann, K.V. : *Physics and the Real World*, CRC Press , Boca Raton, Florida, 1980.

Tolbert, N.E. : *Microbodies—Peroxisomes and Glyoxysomes*, Oxford University Press, London, 1971.

Toraldo, Francia G. : *The Investigation of the Physical World*, Cambridge, New York, 1981.

Verner, J.E. : *Thermal Physics*, Academic Press, New York, 1985.

Vince, Pruce D. : *Elements of Stastical and Thermodynamics Physics*, McGraw Hill, London, 1975.

Walgraif, Daniel : *Introduction to Physics*, McMillan Press, Melbourne, 1969.

Young, H. D. and Others : *University Physics*, Add. Wesley, London, 1996.

Zeevart, J.A. : *Thermal Plasma and New Materials Technology*, McGraw Hill, London, 1996.

Zhukov, L. : *Modern Physics in Twenty first Century, Springer Verlag, Berlin, 1975.*